SINGERS IN THE MARKETPLACE

For
Mark and Tristan

Singers in the Marketplace
The Economics of the Singing Profession

RUTH TOWSE

*Foreword
by
Sir Claus Moser*

CLARENDON PRESS · OXFORD
1993

Oxford University Press, Walton Street, Oxford OX2 6DP
Oxford New York Toronto
Delhi Bombay Calcutta Madras Karachi
Kuala Lumpur Singapore Hong Kong Tokyo
Nairobi Dar es Salaam Cape Town
Melbourne Auckland Madrid
and associated companies in
Berlin Ibadan

Oxford is a trade mark of Oxford University Press

Published in the United States
by Oxford University Press Inc., New York

© Ruth Towse 1993

All rights reserved. No part of this publication may be reproduced, stored in a retrieval system, or transmitted, in any form or by any means, without the prior permission in writing of Oxford University Press. Within the UK, exceptions are allowed in respect of any fair dealing for the purpose of research or private study, or criticism or review, as permitted under the Copyright, Designs and Patents Act, 1988, or in the case of reprographic reproduction in accordance with the terms of the licences issued by the Copyright Licensing Agency. Enquiries concerning reproduction outside these terms and in other countries should be sent to the Rights Department, Oxford University Press, at the address above

This book is sold subject to the condition that it shall not, by way of trade or otherwise, be lent, re-sold, hired out or otherwise circulated without the publisher's prior consent in any form of binding or cover other than that in which it is published and without a similar condition including this condition being imposed on the subsequent purchaser

British Library Cataloguing in Publication Data
Data available

Library of Congress Cataloging-in-Publication Data
Towse, Ruth, 1943–
Singers in the marketplace : the economics of the singing
profession / Ruth Towse.
Includes bibliographical references.
1. Singers—Great Britain—Economic conditions. 2. Singing
teachers—Great Britain—Economic conditions. I. Title.
ML3795.T66   1993   338.4'7782'00941—dc20   93–18832
ISBN 0–19–816347–9

1 3 5 7 9 10 8 6 4 2

Typeset by Graphicraft Typesetters Ltd., Hong Kong

Printed in Great Britain
on acid-free paper by
Bookcraft Ltd
Midsomer Norton
Avon

FOREWORD

by

Sir Claus Moser

Anyone involved in the arts is bound to be interested in economic aspects. As I read this fascinating book, my mind went back to my years as Chairman of the Royal Opera House. Economic and financial matters were never far from our minds as we made artistic decisions. Singers' fees were one such consideration: important of course, though not as dominant as the layman thought. Even so, I used to admire the skill of artists in negotiating contracts and currencies: for me, the model remained the great Melba whose contract with the Metropolitan Opera of New York was said to be a single phrase 'more than anyone else'.

Anyhow, Ruth Towse's approach is more serious than that. She has produced a book that is both important and highly expert, not surprisingly since she is both a singer and an economist. So she brings to her approach the inside knowledge of an artist as well as the disciplined approach of an economist.

What she has done is to link facts about the training of singers with estimates of demand for them and of their earnings. It is often argued that there are too many singers in this world, more than the market can absorb, a view which can clearly influence the world of musical education. But in assessing possible oversupply, one has to study the labour-market itself. This is highly complex, not least, and somewhat surprisingly, because most singers in Britain are freelance. That in itself makes manpower planning impossible: one can't simply establish how many jobs there are for singers and rigorously estimate potential oversupply. Yet the skilled economist can use indirect indicators, such as unemployment and earnings, as a guide to estimates of the 'rate of return' produced by training. This is what Ruth Towse has done.

Her necessarily tentative analysis of 'rates of return' does suggest certain conclusions indicating oversupply, results which are not cheering for budding singers. On the other hand, from the music lover's point of view, perhaps we would rather have too many rather than too few, so as to ensure quality.

In any case, the author does not overstate the strength of her conclusions. What she does, most effectively, is to lay out the issues in simple economic terms, backed by facts and figures on supply and demand, and to show how this labour-market works. Her analysis and policy conclusions will be of value to arts administrators, music colleges, and a wide range of teachers and artists—abroad as well as in Britain. Moreover, since this particular labour-market is unusual, her sensitive analysis will also interest economists generally.

A book so skilfully bridging the world of the arts and economics deserves a warm welcome.

C.A.M.

17 November 1992

PREFACE

THIS is a study of the singing profession in Britain and deals mainly with trained 'classical' singers, performers and teachers. Work took place between 1988 and 1990 at the London School of Economics with a generous grant from the Leverhulme Trust and with some assistance from the Esmee Fairbairn Charitable Trust. The purpose was to study the economic aspects of the training and employment of singers by collecting facts about the way that the labour-market for singers works (or, to be more exact, worked in the research period) and to analyse them using the tools of economic theory. It is an economist's view of the singing profession but one tempered by the experience of years of singing lessons and 'going through the mill'.

But although the outlook is that of an economist, the language is tailored for a much wider audience—arts administrators, those concerned with the education and training of singers (and other types of artist; the principles are the same for all), and singers. Consequently, where economic ideas are used, they are fully explained. However, the book should also be of interest to economists because it deals in great detail with a specific and previously unstudied labour-market.

To many people, an economic study of the singing profession is a contradiction of terms. Many people love to cling to the romantic view of the artist as pure and unsullied by mundane worldly concerns. Yet singers constantly deal with everyday matters ranging from tax returns to train timetables as well as organizing their singing careers. I found many singers to be businesslike, very aware of the financial side of their lives and willing to talk about it. Besides singers and singing-teachers, there are many other people involved with the singing profession—agents, employers of all kinds, administrators of arts organizations and of institutions of higher education. They are also concerned with the economic aspects of the singing profession.

The research undertaken for this book took the form of structured interviews and postal questionnaires. It goes without saying that it could not have been written without the generous co-operation of a large number of people. Questionnaires were sent to singing-teachers, orchestras, opera companies, music colleges, colleges of further and higher education, music centres, private academies, universities and polytechnics, and to foundations and charitable trusts. Someone somewhere in all these organizations filled out a questionnaire and I am most grateful to them. Documentary evidence was also kindly supplied by Equity, the ISM, the Polytechnics' and Colleges' Funding Council, and the DES; in addition, Jack Eliot and Sandy Bailey (Equity), Russell Jones (NFMS), Heather Rosenblatt (ISM), Nicholas Capaldi (South West Arts), James Beirne (Yorkshire Arts), and Phillipa Bird (Greater London Arts Association) gave help and advice; I was also advised by Leon Crickmore and J. R. Williams of Her Majesty's Inspectorate. I interviewed a number of people in music colleges who kindly gave of their time and knowledge: Ian Horsburgh (then at the Royal College of Music), Kenneth Bowen (Royal Academy of Music), Peter Renshaw (Guildhall School of Music and Drama), Sir John Manduell, Joseph Ward and Christopher Underwood (Royal Northern College of Music), Morag Noble (Trinity College), Pamela Bowden (London College of Music), Peter Fletcher and Eileen Price (Welsh College of Music and Drama), Gordon Stewart (Royal Scottish Academy of Music and Drama), Keith Darlington (Birmingham Conservertoire), and Rose Walker (Morley College). On post-college training Richard van Allen and Hugh Lloydd (National Opera Studio), Nancy Evans and John Owen (Britten–Pears School of Advanced Studies) were most helpful and informative. Other singing-teachers kindly gave me their time and advice: Molly Petrie (Secretary of the Association of Teachers of Singing), Geraldine Frank, Mary Makower, Elsie Mayer-Lismann, Ron Murdock, and Professor Wilfred Jochims (Musik Hochschule Cologne).

All the permanent opera companies in Britain were visited and staff interviewed; in addition questionnaires on opera companies were completed by a number of other people whom I did not meet but to whom I am most grateful; I also visited three German opera companies. Thanks go to all of the following for their help and information in interviews and on paper: Paul Findlay and Stephen Laidler (Royal Opera); Jeremy Caulton, Margaret Hall, and Victor Morris (English National Opera); Christine Chibnall, Jim Marsh, and Bruce Budd (Opera North); Richard Mantle (Scottish Opera); Anthony Freud, Sarah Playfair, and Timothy German (Welsh

National Opera); Anthony Whitworth-Jones (Glyndebourne); Claire Foden (Opera 80); Graham Allum and Timothy Dean (Kent Opera); Elaine Padmore (Wexford Opera Festival); Peter Knapp (Travelling Opera); Mary Hill and Anthony Shelley (Abbey Opera); Hilary Griffiths and Claus Henneberg (Cologne Opera); Tim Coleman (Netherlands Opera); Klaus Schultz (Aachen Opera); and Herr Chefdisponent Hauptmann and Herr Broch (Deutsche Oper-am-Rhein). I also interviewed Geoffrey Mitchell and William Robson (BBC), Michael MacLeod (Monteverdi Choir), Terry Edwards (London Sinfonietta Voices), Stephen Hill (Stephen Hill Singers), and Stephen Layton who all gave me a great deal of insight into the concert and sessions world.

I am particularly grateful to a number of singers' agents who supplied me with invaluable (and also anonymous) information: David Sigall and Jonathan Rose (Ingpen and Williams), Diana Mulgan and Robert Rattray (Lies Askonas), Tom Grahame (Harrison Parrott), Robert Gilder (Anglo-Swiss), Neil Dalrymple (Music International), Victoria Smith (Kaye Artists), Caroline Woodfield (John Coast), Mark Hildrew (Ibbs and Tillett), Ron Gonsalves, Jill Davies (Magenta), Michael Kaye (Young Concert Artists' Trust), Catherine Scott (Concert Directory International), Helen Ranger (Ranger Juviler), Gaetano Rea (Portfolio Management), and Angela Halliday. In Germany, Andreas Braun and Herr R. Heissler-Remy.

My greatest debt of gratitude is to the individual singers who kindly agreed to be interviewed; perhaps I should explain here that my aim was to interview singers doing different types of work and at different stages of their careers: Nancy Argenta, Rosemary Ashe, Olivia Blackburn, Keith Bonnington, Tracy Bounden, John Dobson, Faith Elliott, Andrew Forbes-Lane, Carol Hall, Peter Langham Evans, Robert Lloyd, Jane Manning, Harry Nicoll, Anne-Marie Owens, Ian Partridge, Lynda Richardson, Virginia Rushton, Malmfrid Sand, Andrew Shore, Ashley Stafford, Sarah Walker, Alison Wells, and Willard White.

I owe warm thanks to the members of the Steering Committee: Sir Alan Peacock, Professor Meghnad Desai, Noelle Barker, and Hilary Pugh. Thanks go also to my husband Mark Blaug for his continuous support and help and particularly to Diana Moyse for her expert typing of the manuscript.

R. T.

CONTENTS

List of Figures	xii
List of Abbreviations	xiii
Introduction: The Market for Singers	1

PART I. THE SUPPLY OF SINGERS

1. The Provision for Training Classical Singers in Britain	23
2. The Cost of Training Singers	54

PART II. THE DEMAND FOR SINGERS

3. Employment in the Market for Singers	81
4. Factors affecting the Demand for Singers	102

PART III. THE RETURNS TO TRAINING

5. The Earnings of Singers	125
6. The Rate of Return to Training as a Singer	163
Conclusions	191
Appendix 1. Tables	205
Appendix 2. Calculation of the Rate of Return to Training in Singing	224
Appendix 3. The Training and Employment of Singers in Germany	229
Bibliography	243
Index	246

LIST OF FIGURES

4.1	Demand and Supply in the Market for Singers	103
4.2	The Effect of a Shift in Demand on Pay and Employment	103
4.3	The Effect of a Shift in Supply on Pay and Employment	104
4.4	The Effect of Imposing a Minimum Pay Floor on the Market for Singers	104
4.5	The Effect of Imposing a Pay Ceiling on the Market for Singers	105
5.1	Normal Distribution of Income	158
5.2	Skewed Distribution of Income	159
6.1	A Typical Age-Earnings Profile	167
6.2	Comparison of Two Possible Age-Earnings Profiles	167
6.3	Net Earnings Differentials due to Higher Education	169
A2.1	Net Earnings Differentials of Singers	226

LIST OF ABBREVIATIONS

AOTOS	Association of Teachers of Singing
BBC	British Broadcasting Corporation
DES	Department of Education and Science
Equity	British Actors' Equity Association
GBDA	Genossenschaft Deutscher Bühnen-Angehörigen
ISM	Incorporated Society of Musicians
LEA	Local Education Authority
NAB	National Advisory Body for Local Authority Higher Education
NFMS	National Federation of Music Societies
PAYE	Pay as You Earn
PCFC	Polytechnics' and Colleges' Funding Council
UFC	Universities' Funding Council

Introduction
The Market for Singers

The development of the singing profession is closely related to the evolution of the market for singers, the mechanism whereby singers can sell their services for payment in the form of fees or wages. There have been singers since the beginning of civilization but we know little about the singing profession before the advent of the Renaissance in Italy. There were, of course, singers in the Church, but it was with the birth of opera and art song in the Italian princely and ducal courts that singing became a secular profession. That there were from that time on soloists and choristers may be inferred from the operas that were written, for example Monteverdi's *Orfeo*. But though quite a lot is known about the stars, little is known about the rest. Who the chorus singers were, what training they received, and what they were paid is inevitably less well documented than the careers of the star performers. Soloists were trained artists, receiving their training initially in the Church and later from established singers and musicians; singers in the chorus probably had little training.[1] The first professional singers were contracted to work for a patron, usually a prince or duke or other nobleman; they were servants who lived and worked at court or in great houses. They were given their subsistence and, when they pleased their masters, gifts. Later on, as the market for singers developed, they were paid fees or were salaried.

Besides the professional singers, there were amateurs working side by side with them, trained singers who for various reasons, like being a well-born lady, did not join the ranks of the professionals. Not that some well-born ladies did not sing professionally; some did. For many years, however, the singing profession was regarded as being somewhat dubiously respectable. As Adam Smith, the great political economist, wrote in 1776:

[1] For a detailed history of singers and the singing profession, see Christiansen (1984) and Rosselli (1984 and 1991). In the latter, Rosselli writes that chorus singers in Italy in the 18th cent. were usually musically illiterate (see Rosselli 1991: 109).

There are some very agreeable and beautiful talents of which the possession commands a certain sort of admiration; but of which the exercise for the sake of gain is considered, whether from reason or prejudice, as a sort of public prostitution. The pecuniary recompence, therefore, of those who exercise them in this matter, must be sufficient, not only to pay for the time labour and expense of acquiring the talents, but for the discredit which attends the employment of them as the means of subsistence. The exorbitant rewards of players, opera-singers, opera-dancers, &c. are founded upon those two principles; the rarity and beauty of the talents and the discredit of employing them in this matter. It seems absurd at first sight that we should despise their persons, and yet regard their talents with the most profuse liberality. While we do the one, however we must of necessity do the other. Should the public opinion or prejudice ever alter with regard to such occupations, their pecuniary recompence would quickly diminish. (Smith 1976: 124)

Vestiges of both these facets of singing, amateurism and the idea that it is somehow not a respectable profession (like love, highly desirable if freely given but contemptible if sold), persisted right up to the interwar years of this century. Present-day Britain is strong both in its highly respected body of skilled professional singers and in its amateur tradition and there is considerable interaction between the two. On occasion these groups overlap to the extent that it is not an easy task to clearly define the boundaries between them (hence the use of the term 'semi-professional' which usually means that a singer is of professional standard but unpaid).

As the Italian Renaissance spread out to other European countries, the taste for Italian song and opera, and consequently for Italian singers, spread with it. This necessitated the release of those singers who were tied to courts and noble houses and also marked the beginning of an international market for singers in which fees were paid and contracts well established (see Rosselli 1984). Native singers as well as imported Italians began to be employed. At the beginning of the eighteenth century Britain had its own professional singers and has increasingly produced some of the world's finest singers. Eventually singers of all nationalities were working on an international circuit covering Europe from Britain to Russia in the eighteenth and early nineteenth centuries, and by the middle of the nineteenth century this included America and subsequently South America. Nowadays the market for singers is truly worldwide, including Asia as well as Australia in addition to Europe and the Americas; this world market supplies singers of many nationalities, many countries throughout the world now having music colleges and conservatoires training singers as well as demanding their services for work in concerts, recitals, opera, and musicals.

With the breakdown of ducal and princely courts in Italy, Germany, and elsewhere, economic and political power was transferred to elected state authorities. With the growth of capitalism and market economies, singers were free to find work in the marketplace and so the labour-market for singers began to operate. The opera-houses came under the newly formed state and municipal governments with state patronage in the form of subsidies often replacing the private patronage of princes and nobles. Occasionally privately owned opera companies emerged, such as the Bayreuth and Glyndebourne Festival Operas. The present-day organization of contractual arrangements between singers and opera managements still reflects the history of these changes in ownership. In Britain there was practically no state subsidy of opera until 1946 and opera had a rather up and down history until it was regularly financed by Arts Council grants. This is reflected in the fact that the labour-market for singers in Britain is still largely dominated by freelance arrangements, even in opera. In oratorio, the great nineteenth-century tradition always called for freelance soloists. By contrast, most German opera-houses have singers on regular contracts and salaries, reflecting the different historical developments there. Understanding the economic logic behind the way that the labour-market for professional singers works is an essential part of the subject of this book.[2] What is meant by this is discussed later on in this chapter.

The Picture in Britain in 1990

The growth of state subsidy for music and opera after the Second World War made considerable changes to the singing profession in Britain. On the demand side there was a sustained growth in the provision of music and opera, such that Britain in 1990 had five permanent full-time opera companies, the Royal Opera, the

---

[2] This brief sketch of the origins of the market for singers may whet some readers' appetite for an historical treatment of the development of the music profession in Britain. The definitive book on this is Ehrlich (1985). However, he has much more to say on instrumental musicians than on singers.

English National Opera, Opera North, Welsh National Opera, and Scottish Opera; all these companies were subsidized by the Arts Council of Great Britain (the Welsh and Scottish companies getting their grants through the Welsh Arts Council and the Scottish Arts Council). In addition, there were several well-established opera companies with a regular season for several weeks or months in the year; Glyndebourne Festival Opera, which was privately financed, Glyndebourne Touring Opera, subsidized by the Arts Council; Opera 80, which was created by the Arts Council and financed to tour Britain; Buxton Festival Opera, Opera Northern Ireland, City of Birmingham Touring Opera, which received some public subsidy; Pavilion Opera, and Travelling Opera, both privately financed. Besides these companies there were many smaller ones, some of which had been in existence for quite long periods and others of which came and went. During the 1980s three of the bigger companies ceased to be supported by public funds; Kent Opera, New Sadlers' Wells Opera, and D'Oyly Carte Opera.

Besides the opera companies there were many other musical organizations which employed singers, particularly the professional orchestras and the amateur performing societies, which employed professional singers as soloists in concerts and oratorio. Festivals provided important opportunities for recitals and concerts involving singers. The BBC ran its own professional choir, the BBC Singers, and promoted recitals by singers, broadcasting performances by other promoters, and some oratorio and opera; in addition it also ran the Proms, offering many opportunites for professional singers in all fields.

The BBC and other television companies also occasionally promoted opera. The Church continued to offer work for professional singers, some of whom were employed full time, though most just worked on Sundays and for weddings and other special occasions.

With all this work available, surprisingly little was of a regular nature. The five permanent opera companies had permanent, full-time choruses and several of them had some regular principals in their companies, mainly the English National Opera and the Royal Opera. Most principal roles, however, were performed by freelance singers employed on short-term contracts. These companies along with the BBC Singers and some cathedrals were the only employers offering full-time contracted work to singers. The other opera companies employed the chorus full time for the duration of their seasons and in a few cases this could amount to full-time work, e.g. being in the chorus of both the summer season of Glyndebourne Festival Opera and then doing the tour with Glyndebourne Touring Opera, but, again, principal roles were usually taken by freelance singers. In 1990 jobs which carried regular contracts represented only a small part of the total work available and all other singers—opera principals, soloists in oratorio and recitals, concert and sessions singers—were freelance. Even those singers who were regularly associated with certain choirs or choruses were actually hired by the session.

On the supply side, the growth of state subsidy since 1945, this time not to the arts but to education, increasingly led to the development of courses in institutions of higher education offering training to students of singing.[3] Courses for singers had existed for some time in the music colleges and conservatories but they gradually developed in universities, polytechnics, Colleges of Higher Education, and later in Colleges of Further Education as well (though the latter strictly speaking should be not included as institutions of higher education). There were also private academies and other organizations offering courses and tuition at all levels. Another form of training was provided through choral scholarships, mainly in the Oxford and Cambridge colleges but also at some other universities. Advanced and postgraduate training in singing and opera was provided in the music colleges and by the specialist National Opera Studio. There was also a well-established private market for singing-teachers in Britain. Although the majority of trained singers had attended a music college, if not as undergraduates then as postgraduates, it was nevertheless not a requirement of any employer that they had done. The demand for singers did not depend on paper qualifications.

Indeed, the possession of a qualification in singing neither ensured that a singer was able to obtain work in the singing profession nor that they would necessarily wish to. In this respect, there were no entry qualifications to the singing profession and it was therefore hard to know exactly how to define the singing profession; it is a profession in the sociological sense that trained singers have a professional group identity and membership but there are no

---

[3] In Britain, provision for higher education in the arts was part of the higher education sector, not the arts sector. There was one exception: the National Opera Studio was subsidized by the Arts Council of Great Britain.

specific economic criteria, such as doing paid work, or supporting oneself financially by singing, that provide a satisfactory definition (see Towse 1992c). The annual flow of supply of singers was only loosely connected to students obtaining a qualification as a singer in any one year. In the labour-market for singers in the 1980s nothing hinged on what type of qualification a singer had received; it might help in the very first instance to get an audition but potential employers—the opera-houses, conductors, fixers, promoters, and the like—went entirely by what they heard at auditions or performances. Thus, the supply of singers could not simply be equated with the output of people with qualifications in singing from institutions of higher education.

Reference has been made several times to the term 'the labour-market for singers' and it is now time to explain what is meant by that. This is discussed first as an economic concept and later the way that the market works in practice in Britain is outlined.

What is a Labour-market?

The term 'labour-market' is used to register the fact that exchange takes place between those who want to employ the services of others who are offering their time and skills; payment for the use of those services may take the form of fees, wages, or salaries. A fee is the money paid for a specific service rendered, usually one performance of a particular work; wages and salaries are paid for a specific period of time (weeks or months) over which work of a particular kind has been undertaken. The exchange of services for payments is regulated by contracts, which specify the type and amount of work involved and the rate of pay. The contract may be of a standard form, for instance, that negotiated by a trades union or professional organization, or it may be agreed on an *ad hoc* basis by the individual participants. 'Conditions of employment' relate to the exact nature of the job to be undertaken and how it is to be organized; they are agreed in the contract. The term 'conditions in the labour-market' is also used, meaning the institutional or economic conditions which affect the way the labour-market works. On occasion, a labour-market might be very specifically defined to include only one section of the whole market, for example contemporary music or Early music. However the term is used, it will always denote the fact that what is under consideration is the demand for and supply of labour.

There are a number of standard questions that are asked about any labour-market. Does the labour-market work freely or is it subject to restrictions? For example, can any one join the labour force or are there qualifying examinations which need to be passed, or is it necessary to be a member of a professional society or trades union? Is it competitive or are there elements of monopoly such as a single employer or a tight-knit group of employers? How is the labour-market organized—how do people actually look around for work or workers? Can workers easily move from one part of the labour-market to another or are they 'segmented', that is boxed into one section of it? Is there sexual and/or racial discrimination? All these questions are really about how supply and demand work in the labour-market to determine payments and the level of employment. This is not the only approach to be taken to labour-markets; sociologists, for example, also look at labour-markets and ask questions about status or social class; some of the questions overlap and sociologists' and economists' ideas sometimes interact. Several of the surveys of arts labour-markets that have been done are by sociologists. But what is central to the economists' approach is the notion of supply and demand and how participants in the labour-market respond to economic incentives, such as rates of pay and costs.

In any market that works freely, the interaction of supply and demand will determine the price at which exchange takes place. The wage rate or fee is the price of a particular unit of work in a labour-market. How many hours of labour are offered, that is, the supply of labour, will depend upon the amount of the wage rate or fee. It is usually thought to be the case that the higher the payment, the more hours that will be offered to the labour-market, either through existing singers working harder or by more people being tempted into the profession. There is evidence that singers do respond to economic incentives; one hears singers say that it is not worth their while to sing something at a fee of less than £x: you also hear of people who decide they can 'take the risk' of entering the profession full time when their fees start to rise. This is not to say that the only reason they sing is for money—that is much too simplistic. What is the case is that the decision as to exactly what to sing, where and how much work to undertake, at least for a fairly sought-after singer, will often be decided on the basis of different payments offered. For example, more singers work in opera because they can earn more in opera than in other branches of singing. But there are a few singers who have no idea how much they earn and leave fees and the like entirely to their agents; at the other end of the line, there are many struggling singers who, so far from having to choose between jobs, can hardly get enough paid work to keep going. The supply of singers' services does respond, albeit not exclusively, to changes in payments and this is true both at the individual level and of the whole profession taken together. Equally, on the demand side concert promoters, opera-house management, orchestras, Festival managements—let us for simplicity call them employers—will generally speaking prefer to pay less rather than more for singers' services and so demand is affected by the amount of the fee; if fees rise, demand will fall.

A detailed analysis of the supply and demand of singers is discussed in several places in this book and the question of whether the labour-market for singers works like other labour-markets is raised. Assuming for the moment that it does, what does that tell us? First of all, if supply is greater than demand unemployment will result; that is a situation of 'excess supply' and there has been much discussion as to whether there is an excess supply, or oversupply, of singers being trained in colleges in Britain. There is a problem of how unemployment is to be measured in a situation in which there are no norms about hours or the age of retirement—what is meant by 'full employment' in the market for singers, where the bulk of the profession is working freelance? That apart, if supply increases with demand remaining the same, more people will be chasing the same number of jobs and unemployment will result. But if the rate of payment were to fall, demand would increase and there would be more work available. Therefore the interaction of supply and demand determine the level of employment as well as the rate of pay that balances out these opposing forces. The rate of payment regulates the market; if there is a shortage of singers, payments will rise and so more singers will enter the profession in the long run, and if there is a surplus, payments will fall and some will eventually leave. That alters the balance of supply and demand and the process continues until the right balance is found.

So the factors affecting the demand and supply of singers determine the willingness of employers to hire singers at a certain fee or rate of pay and the willingness of singers to work at it. These factors can be due to events outside the labour-market; on the demand side, the level of arts subsidy to orchestras, opera companies and suchlike affects their budgets; so does the rate of inflation. If the value of their budgets falls then they have to cut back, either by spending less on singers or giving fewer performances. On the other hand, the rising popularity of opera with audiences means that more money is coming in through the box office and this allows opera companies to hire more and better singers, to do operas with bigger production costs, and so on. On the supply side, if there were changes in the funding of higher education, places in music colleges and other institutions which offer training to singers might be cut, or fees and other costs of training might rise, causing fewer students to study singing. What they could earn in other jobs or professions might also influence the supply of singers; many singers leave the profession because they say they cannot afford it, meaning they could do better elsewhere. Many singers work abroad because pay is better and there is more work available than in Britain. These factors affect the supply and demand of singers and hence levels of pay and employment.

There are other factors besides state subsidy that affect the free operation of labour-markets; trade unions and professional associations intervene in the market process and supplant the interaction of supply and demand by the collective bargaining of rates of pay. In the labour-market for singers, Equity and the ISM set minimum rates of payment—fees or wage rates. Equity minimum rates for opera singers (principals and choristers), concert and sessions singers, church singers, singers in musicals, etc. are negotiated annually with the respective groups of employers—the opera companies, record, radio and TV companies, the BBC, Theatre Managers' Association, and the rest. These rates of pay replace individually bargained rates for in this type of work, which would otherwise be set by supply and demand. The ISM sets a minimum recommended rate for solo work and teaching. However, the forces of supply and demand strongly influence the parties concerned in the collective bargaining process. If the minimum rate is out of line with the rate of payment that the market would bring about, unemployment could result (because supply is greater than demand) for some singers, while those in work would earn more per hour or per session than they might otherwise have done. Trade unions often have to make the choice between a wider membership and protecting the earning power of those who can expect to get work; some solve this tension by effectively controlling the supply of labour and limiting membership. The 'closed shop', a situation in which employers make binding agreements only to hire union members and pay the agreed rates of pay became illegal in Britain in the 1980s; even so, Equity has maintained its agreements with employers voluntarily and the change in the law did not lead to substantial changes in practice. Equity has limited full membership only to performers in the entertainment industry who have obtained a certain amount of work at its minimum rates; generally, it has not been too difficult for classically trained singers to join Equity.

In order to analyse how the labour-market for singers works it is necessary to make generalizations. To do that it is necessary to show that, by and large, people—singers and employers—will react in the same way to market forces and incentives. If all that could be said is that everyone is different, economics or sociology would have little to say. There could be no theory of demand if opera-houses or conductors treated each and every singer as an individual case. Research undertaken for this book showed that generalizations could be made but it also showed that there was considerable variation that made generalizations difficult in certain respects. All this is not to say that singers' individuality or an orchestra's or opera company's policy does not matter or does not need to be taken into account; it is simply to say that many individuals and organizations do behave in the same way when faced with particular economic decisions. Supply and demand involve decisions—how many hours or sessions or performances to work for a certain rate of pay or fee; how many hours or sessions or performances to hire or book or contract for at a certain rate of pay or fee. All these individual supply decisions can be aggregated to yield the total supply to the market and equally all individual demand decisions taken together constitute market demand.

The Organization of the Market for Singers In Britain

On the supply side, the starting-point is formal training. For most classical singers in the 1980s, formal training in a music college or conservatoire or any of the other institutions of higher education, even if it included a specialized postgraduate course in, say, opera, was only the beginning, and relatively few of them immediately got full-time or even part-time employment as singers. Many singers are simply not yet ready for professional work when they leave college. The National Opera Studio took twelve singers a year for its one-year advanced training course and a few of them came straight from college, but most had had some work experience before they entered. Those singers who had acquired sufficient vocal training and other professional expertise, such as stage experience, might soon find work in the seasonal opera choruses, particularly the two Glyndebourne choruses; but the opera choruses of the permanent opera companies, which anyway had only two or three vacancies a year, typically require considerable vocal stamina and make considerable demands on singers. Welsh National Opera, Scottish Opera, and Opera North tended to favour younger singers and English National Opera and the Royal Opera certainly considered recent graduates, but it is fair to say that very few would be suitable. A few very promising young soloists were taken on by Glyndebourne and the Royal Opera, which had young singers' schemes, and could be given work as covers (understudies) and small parts. Some recently trained singers could be accepted for *ad hoc* work in extra choruses. It has to be understood that in the context of the singing profession, 'young singers' means people in their mid- to late-twenties; many would-be soloists spend six years in formal training.

Singers do not have to decide at the outset of their careers whether they wish to aim for a solo career or to opt for chorus and other group work but they do have to be careful not to do work that would render that choice impossible later on. Although there used to be a demarcation line between chorus and solo work, one that was difficult to cross, that is no longer quite so much the case. But if a singer does not have the scope to develop his or her voice individually, it may be very difficult indeed to do that later on. One has heard stories of singers who have spent a few years in sessions work or in opera choruses needing to spend a year or so to retrain the voice to develop an individual sound. Employers and agents do not refuse to consider singers for solo work who had worked extensively in groups, but experience has shown that it can be difficult to make the transition. That said, there have been quite a few cases of star singers who have worked their way up to the top of the profession from a beginning in chorus, sessions work, or musicals.

There are no actual barriers to entry as a soloist other than outstanding ability.

Most singers, therefore, on leaving college faced several years of freelance work, sessions and church work, odd solo jobs with choral societies or amateur opera companies, the odd lunch-time recital, and suchlike. However, there were some employers who preferred young singers. This was usually the case with the BBC Singers, which besides having a permanent ensemble also had quite a number of other singers on its roster whom it booked for *ad hoc* work. It was also true of Early music groups, which wanted singers who did not have much vibrato in the voice; these groups in fact often favoured choral scholars who had acquired considerable training in learning to blend their voices with others' and who had a lot of experience of working with different conductors, of sight-singing, and even of recording. Music colleges tended to train students as soloists and they offered little of this type of experience.

So what did most young singers do? The answer is they found a variety of ways of supporting themselves while continuing to have singing lessons and building up their repertoire. They might be lucky in getting bursaries from private charitable foundations, or they may have won prize money in competitions, or got help from parents, and they did all sorts of part-time jobs. Some taught, either singing and/or piano. Anything and everything went into keeping themselves going and gaining what performing experience they could.

At this stage singers often attended masterclasses, for example those run by the Aldeburgh Foundation, or went to summer schools, some of which were highly professional and specialized, such as Clonter Opera Farm. In London they joined specialized Adult Education courses, such as Morley College Opera or Abbey Opera — there was quite a number of these (of varying quality). These activities, which were far from cheap, gave singers experience in standing in front of audiences and in developing the voice and stage personality. Much of a singer's real training comes from on-the-job training and experience. But besides gaining experience, they were also getting themselves known to potential employers, who keep their eyes on all the above-mentioned activities in their search for new singers.

One might at this point ask what the music colleges did to help students with the transition from college to employment. At one level there is not much that they could do, apart from giving them a good grounding in the necessary skills. To some extent colleges contented themselves with this and they have frequently been criticized for not doing more to teach students about how to find their way in the world of work.[4] Some improvements have been made, though many agents and employers complained that many recent graduates still had no idea, even how to write a proper letter of application or curriculum vitae. It seemed that teachers were often unable to offer proper advice or did not want to take out time from what they regarded as their real work of teaching singing; another (possibly related) reason was that students were very uninterested in the talks that colleges arranged to be given by people working the profession. Though colleges did advise students on how to audition, and on repertoire, many students still showed little understanding when the time came. Common mistakes, apparently, were singing arias from operas of which they had never bothered to find out the plot; singing Lieder at opera auditions, not having legible music for the pianist, being late, etc. (see Legge 1988: 54–9).

Putting such problems aside—and music colleges cannot be blamed for all of them—what were the positive contributions that colleges made to help students get work? The most obvious was the opera productions and concerts that they mounted which provided a showcase. These were put on at least once a year in each music college and those of the bigger ones were attended (and reviewed by) critics from national newspapers, *Opera* magazine, and so on, as well as by agents and other talent scouts. Particularly in the world of opera, there was a wide-ranging grapevine monitoring talented singers almost from the cradle to the grave! (It should be said that because the singing profession is essentially relatively small and there is a great deal of movement within it, many jobs are obtained through word of mouth rather than by advertisement; if you are not 'on the circuit' it can sometimes be very difficult to find out about job opportunities.) The National Opera Studio had an annual showcase performance of scenes from operas which were hand-picked to show off the singers in their chosen repertoire. As with the music college productions, these were always attended by critics, agents, opera company officials concerned with casting, and

---

[4] See e.g. the report of the Calouste Gulbenkian Foundation (Vaizey 1978: 115–17). There have been a number of other important reports on the training of musicians. They are discussed in some detail in Chapter 1.

other such persons. In addition, the Studio arranged a concert tour and sent its singers to work with the orchestras and conductors of all the main British opera companies. It also invited foreign opera company officials to private showings. By the time a singer left the Studio, he or she certainly had had the chance to perform to a wide selection of employers.

Music colleges also helped students who were of a suitable standard to get small solo jobs, say as soloists with church choirs or with amateur performing societies doing oratorio. They posted notices of these jobs on notice-boards and some colleges had an office which dealt with such requests. Some also organized concerts in hospitals or old peoples' homes and the like, which gave students the opportunity to perform before an audience.

Colleges also arranged auditions for those singers it considered ready for professional work, particularly with agents. This brings us to another aspect of how the market for singers works, the role of agents. Agents are as old as the singing profession itself (Rosselli 1984). Agents and managers are middlemen who act on behalf of singers to find them work and negotiate contracts. They do this through their generally wide knowledge of the singing profession; they keep in touch regularly with opera companies to find out what operas they are going to cast, and opera companies get in touch with them for information about suitable singers. Similarly, they are in touch with conductors and orchestras for concert work and oratorio, with the amateur performing societies and music clubs and with festivals and other promoters. Once they have taken a singer on, they charge a commission, 10–15 per cent of the fee obtained for the singer in 1990. Being profit-making entrepreneurs, agents have always come in for a great deal of mistrust and, no doubt, in some cases this is not unjustified. However, there are agents and managers who have a responsible attitude to the development of their singers' careers—indeed, it is in their financial interest to do so. In 1990 there was one non-profit-making agency in Britain, the Young Concert Artists Trust, which helped a few singers (they had two on their books that year) at the start of their professional careers and then passed them on to other agents.

Generally speaking, agents in Britain were not interested in singers until they had quite a lot of performing experience and had shown promise of being able to sustain a reasonably long career, commanding good fees. Agents did take on one or two very promising young singers at the early stages of their careers, even while they were still at college, whom they would nurse along until they could begin to work regularly. However, because a good agent cannot generally take on a large number of singers and give them individual attention, agents only accept those that they expect to make a successful career—otherwise they would not earn much from them, which they need to do to survive. Most established singers have agents. A busy singer would not have the time to deal with all the arrangements that agents and managers make—performance and rehearsal dates, travel arrangements, and so on—and they take away from singers the burden of having to negotiate fees, a task that many singers do not like and also are often ill-equipped to do. From the economic point of view, agents fulfil a very important function. They cut down the costs singers have in looking for work and getting information about fees. These are known as search and information costs. Agents also, though this is not their purpose, cut the search and information costs of employers as well. But even more importantly, agents have information about the fees of other singers and so have a sense of the going market rate of performance fees. Agents therefore aid the competitive process in the labour-market for singers. But though nearly all successful singers have an agent, this does not mean that a singer cannot get work without one. Enquiries showed that in Britain it was open to any singer to approach an opera company or a conductor for an audition.

One final question should be raised here: is there a career structure in the singing profession? What is normally meant by a career structure is an understood progression from one stage of work to another, usually with an earnings pattern to match. As the market for singers, at least in 1990, was largely dominated by freelance arrangements, there was no set structure of work and certainly no set salary structure through which singers were promoted. In the permanent opera choruses, where singers were salaried, there was no real career path either, though singers might get paid a bit more after two or three years' employment; in some of the choruses, all singers were paid the same however long they had been working with the company. In most other choral work and in concert and sessions work, only the Equity minimum was paid, which was the same regardless of age or experience. Principals' salaries for those regularly contracted to companies were individually negotiated and followed no set pattern. For freelance singers, it is just a matter of how well they could do in the marketplace. But this is not to say that there was no sense of progress about singers' careers. In general, there was the sense that solo work conferred a higher status than chorus or group work and international work conferred a higher status than all-British work and that was usually reflected in fees. Singers all had to learn individually, and often painfully, what their rating was and what kind of career they could hope to make. It was very much a case of 'many are called but few chosen' and many suffered huge disappointment and financial difficulties. Because it cost a great deal to train a singer, and much of the cost was borne one way or another by singers themselves, and because of the uncertainty of the singing business, many a singer has wondered whether they made the right career decision.

Specialization in the Market for Singers

At one time, singers more or less had to choose whether to specialize in opera, recital work, or oratorio but this has now become increasingly less the case. A well-trained singer in 1990 was able to work in all these fields and would have studied the whole range of repertoire as part of his or her training. That is not to say that singers do not specialize nor that every voice is suitable for every operatic role or every style or period of music. It will always be the case that some singers are more comfortable with one type of repertory rather than another and this can lead to shortages of one type of voice, or of singers who can sing one area of repertoire e.g. Wagnerian tenors. There were no 'rules' about this which prevented singers from moving between the different branches of singing, but the effect of specialization could be strong and in some cases lead to a clear division of labour. Many opera singers cannot easily sing Early music and Early music singers might not have big enough voices for the larger opera-houses. This leads to subdivision of the market according to specialization. Another area of specialization is related to the ability to sight-sing. Though all singers are taught sight-singing during their training, at least in music colleges, by no means all of them acquire facility in it. All sessions work and much concert work requires a very high standard of almost flawless sight-singing. This raises a very difficult issue to which nobody seems to really know the answer. Some teachers and conductors are of the opinion that sight-singing cannot really be taught, and certainly singers who excel at it have often learnt it through their very earliest musical training either at school or in singing with choirs or through learning to play an instrument. Yet people with great singing voices may not have had an early musical training; very few know what their voice will be until they are in their late adolescence or adulthood, by when it seems to be difficult to achieve real facility in musical skills. The result is that there are some great singers, who, though vocally and musically gifted in other respects, have only a minimum competence with basic musical skills. (There were reputed to be international star singers who even in 1990 could not read music at all, but this might be apocryphal.) In opera, and recitals, where repertoire has to be learned by heart, this need not be a problem, since principal singers work with accompanists and opera coaches (repetiteurs—usually pianists—who teach repertory to singers) and choruses had rehearsal time in which to learn words and music. But in sessions work this would be impossibly time-consuming and expensive, and unless singers can sight-sing they cannot work in this area. A further requirement in some branches of contemporary music is that singers have perfect or as-good-as-perfect relative pitch.

Singers, therefore, like other workers, specialize in areas where they have a comparative advantage. Most voices fall naturally into one voice category or another—soprano, mezzo-soprano, countertenor, tenor, baritone, bass—while others are more difficult to classify or change with training or age. Some singers, therefore, have scope for choosing what they specialize in while others are limited by nature. Some singers may choose to specialize in a particular field for economic reasons, for example there are strong economic incentives for tenors to take on Heldentenor roles even if the voice is not naturally dramatic. Luckily in Britain this has not been too much of an issue, but in German opera-houses a singer has long been placed in a detailed 'Fach' with a contract requiring that they sing any role in that Fach (see Appendix 3; Legge 1988).

On the demand side, the choice of voice types may vary too; the most obvious contemporary example of this is the substitution of countertenors for women's voices in Baroque and earlier music but there are other examples—the composer in Strauss's *Ariadne auf Naxos* or Octavian in *Rosenkavalier* may be sung by sopranos or mezzos; Rosina in Rossini's *Barber of Seville* and Bizet's *Carmen*, prized mezzo roles, were sung by Callas; Gluck's Orfeo has been sung by tenors, counter-tenors, contraltos, or mezzos. Gluck's *Orfeo* is a case of an opera whose title role may be sung by various voice types, depending on which version is selected; this choice may be influenced by economic as well as artistic considerations. Tradition may also determine the choice of which voice type to cast. But substitution is possible, at least in Britain, and where a choice may be made, again it may be subject to economic influences. Within certain bounds then, singers are free to specialize, but they need not get trapped in doing so. If they wish, most singers could move between one section of the labour-market and another.

The International Market for Singers

Singing is an international art and there has been considerable movement of singers in an international labour-market for several centuries. All trained singers in Britain are able to sing in several languages, particularly German, Italian, and French, and many can sing in Russian, Spanish, and Czech. Many British singers work abroad, some for long periods of time on regular contracts (Germany being a big market for opera singers) but also for one-off runs of performances or recitals. Although this study was completed before 1992 and the single market, there was already freedom of movement for musicians.

The international market has two effects on the British market for singers; it offers considerably expanded employment opportunites and it pays well. In the 1980s fees were considerably higher abroad than in Britain and though living expenses could be higher as well, many singers worked abroad because it was financially attractive. British singers, both choral singers and soloists, are well thought-of internationally because they are well trained and have high professional standards. (What seemed to be meant by the phrase 'high professional standards' was knowing the music before rehearsals started—or being able to sing at sight, something that apparently has evaded some Continental singers—attending rehearsals, being co-operative, and not cancelling performances at will.) At some stage most successful singers seem to have worked abroad and many singers made their name abroad before they achieved real success at home. This type of competition generally affects the supply of singers willing to work in Britain and also causes fees to rise due to international competition. The fees of international singers tend

INTRODUCTION: THE MARKET FOR SINGERS

to equalize over all the countries in which they work so that costs rise for the opera-houses and other employers who hire international singers. Thus the British market has been and continues to be influenced by the international market for singers.

In this introduction, the many facets of the singing profession and the labour-market for singers have been sketched. In the rest of the book, the economic aspects of the training, employment, and earnings of singers are analysed in detail. Chapter 1 presents the results of a number of surveys on training, which have produced the first comprehensive estimates of the number of singers being trained in Britain in 1988/9. Chapter 2 looks at the costs of training —how much it costs the individual trainee as well as the government. These two chapters deal with the supply of singers. In Chapter 3, the results of my enquiries about employment for classically trained singers in Britain are presented and Chapter 4 discusses what influences the demand for singers. In Chapter 5, singers' earnings are dealt with and a considerable amount of data assembled and analysed. Chapter 6 puts together the findings on the cost of training and the return to it from earnings and considers whether it is financially worthwhile to become a singer; the issue is discussed from both the individual and societal point of view, in the relation to calculations, albeit hypothetical ones, of the rate of return to training as a singer. The conclusion contains a summary of the main results of this study and draws policy implications from them.

Throughout, the focus is on supply, demand, and how the market works. The main economic ideas have already been introduced and they do not become substantially more difficult in subsequent chapters. All the tables and calculations have been put into appendices and the book can be read straight through without having to scrutinize them in detail. Appendix 3 contains a brief description of training and employment of singers in Germany, which provides an interesting contrast to the labour-market for singers in Britain.

Part I
The Supply of Singers

1
The Provision for Training Classical Singers in Britain

The supply of labour, in economic terms, is the number of people willing to work, or the hours of work that people are willing to offer, at various wage rates. This notion of supply is very difficult to measure, however, and as a result, other indicators are sought. One such is the size of the labour force (the stock of available workers), for example how many professional singers there are. Another indicator is the flow of new entrants to an occupation or profession; this is often measured by the output of relevant training courses, for example singers leaving music college.

Using this measure of the flow of supply assumes three things: first, that all trainees will wish to enter the profession; second, that qualifications are accepted as proof of professional competence by employers and third, that there are set ways of qualifying to enter a profession, for example by attending an institution of higher education (college, university, or polytechnic). These assumptions underlie the manpower planning approach to the supply of training personnel of all kinds, and this was the approach implicitly adopted by John Vaizey in his important report *Training Musicians*, commissioned and published by the Gulbenkian Foundation in 1978 (Vaizey 1978). Despite its date, it is the most recent enquiry that has been conducted into all aspects of musical training in Britain. But Vaizey's adoption of the manpower planning approach was misconceived in the context of the music profession and, in particular, in relation to the singing profession.

The Manpower Planning Approach

Manpower planning claims to answer the question how many new entrants to a profession or job should be trained in order to provide sufficient trained personnel for future 'needs of the profession', or, to an economist, demand. This approach assumes on the one hand that the demand for people with certain skills can be forecast some years ahead, and on the other that, having received a certain training, students will both be able to enter their chosen profession and will choose to do so. While this approach may be appropriate to licensed professions where the government controls demand, for example as with teachers and doctors, it certainly is not appropriate to musicians.

The demand for musicians is subject to market forces; while demand is to some extent affected by the amount of government subsidy available to the arts, that is a very far cry from the situation, say, in school teaching, where the demand for teachers by the public sector is guaranteed and can be forecast from birth rates and other such indicators of future demand. Where there is a thriving freelance market, as there is for musicians, demand cannot be tied to posts or to a set number of jobs. Demand can rise without requiring a greater supply, since the same people can just do more work; for example, though the number of opera performances has risen, this has largely led to more performances of the same operas, making more work for the existing cast, rather than for more singers. Nor is the supply of new entrants to the music profession a straightforward matter either. The supply of musicians is not determined only by the number of places in music colleges and other institutions of higher education; it is determined on the one hand by how many trainees want to try to make a career in music given the financial rewards on offer and, on the other, by the number of entrants to the profession offering what employers deem to be acceptable skills and experience. In the singing profession in particular, a degree or diploma is not regarded as proof of their attainment. There are many routes into the singing profession and, though music colleges play an important part in training singers, they offer only one way of acquiring the necessary skills. Furthermore, singers who have trained at music college often need further training time before they are ready to start a career; thus the output of singers from music colleges and other institutions of higher education is not the same thing as the number of new entrants to the singing profession. None of these difficulties are taken into account by the manpower planning approach. Unfortunately, there is no simple way of answering the question 'how many musicians should be trained?' as this book shows in subsequent chapters. In fact, there is not even a simple way to answer the question of how many singers are being trained in Britain at any one time, or what constitutes proper training for singers.

The Training of Singers

Where should one start in discussing the training of singers? Most singers will say that they cannot remember a time when they did not sing—at home, in school choirs and concerts, in church, and so on. Some singers, chiefly (though not exclusively) men, went to choir schools as boys and so received formal voice training from an early age. The significance of all these early experiences cannot be overestimated; besides laying the foundation of musical skills for those who will subsequently become professional instrumentalists and singers, they also form the basis of the amateur choral tradition, which is enormously important to the British singing profession.

In general, serious vocal training does not begin until late adolescence. There are some early developers who can begin younger (Malibran began her professional career at the age of 13, Jenny Lind at 10);[1] there are also many examples of late developers, people who did not even know they had a voice until they were in their thirties; there are a few examples of singers who made a professional début when over the age of 40. There are many exceptions to the general rule. However, there is a general rule and that is that most singers start formal vocal training at around the age of 18 to 20. This is undoubtedly largely due to the fact that this is the usual age at which higher education begins in Britain and it is not the ideal age for commencing vocal training; heads of singing in the music colleges would prefer 20 as the age of intake, and indeed, a few years ago agreed between themselves not to take students until that age: although the arrangement fell through because of the intense

---

[1] See Christiansen (1984) for the biographies of a number of famous prima donnas. Another book which gives some details of the early training of singers of this and the last generation is Hines (1983). Apart from these books, there are hundreds of biographies and autobiographies of singers; there are also biographical details on singers in New Grove (1980) and in other similar dictionaries—all too numerous to cite. They testify to the considerable variation in the length and type of famous singers' studies as well as in the age at which they made their professional début. The inclusion of music colleges in the system of state-subsidised higher education has imposed more uniformity on the training offered to recent generations of singers, both in Britain and elsewhere.

competition between the music colleges for good students (and also apprehension as to what applicants would otherwise do), heads of singing nevertheless prefer students to enter college at 20. Even at 20 some students, particularly men, are not really ready to start intensive training and colleges admit students throughout their twenties. Nowadays, many would-be singers go to do a degree in a university or polytechnic first (in a range of subjects, not only music) and then go to a music college afterwards.

At this point the question must be raised as to what sort of training are we talking about? After all, many people sing regularly and often without any formal training—singing is a natural thing to do. But although it is natural, singers need to develop the vocal stamina to sing for several hours a day, day in day out (running is natural but everyone could not be an athlete—and singers are in many respects like athletes). That is the essence of classical vocal training. People do sing without training, pop stars, for example, but they do not need to project their voices in large spaces without microphones—nor do many of them have a long career. But it is not only opera or concert soloists who need to be trained; choral singers need training too; they may not have to project their voices in the same way as soloists but they have to blend them together. So the question as to what is an appropriate training to some extent depends upon which branch of the singing profession a student is aiming for. This, however, is complicated by the fact that there is no simple equation between training and the demands of the labour market.

In the 1980s there was no single standard means of training; there were many routes into the singing profession and a range of provision for training singers. All the music colleges (or conservatories) offered specialized courses for singers; many universities, polytechnics, and Colleges of Higher Education also offered courses with singing as a special option. In addition, there were around two hundred choral scholarships in British universities. These courses formed part of the provision in the higher education sector. Singing lessons were also offered as part of a course in Colleges of Further Education and in Local Authority Music Centres and in some specialist schools. There was well-established and extensive provision for singing lessons at all levels, from beginners to advanced professionals, with private singing-teachers. Finally, specialized short courses were offered by a variety of public and private organizations

THE PROVISION FOR TRAINING SINGERS

for all levels and age groups of singers, amateur and professional. Research for this book included a survey of all these types of provision and the type of training they offered is discussed later in this chapter.

What constitutes a proper training for professional singers is a difficult question. It is argued here that it is the labour-market ultimately that answers that question. This book is about the economic aspects of training and it would be inappropriate to comment on questions of the quality or content of training courses for singers. However, there have been several expert reports over the last twenty-five years dealing with qualitative issues in the training of professional singers, though they mostly concentrated on musicians in general and only discussed singers *en passant*. None investigated the costs of training and though they looked at employment, that was never given a proper economic treatment. While the training for singers and instrumentalists seems to call for similar treatment within higher education, the long-run economic aspects of their training are in fact very different, mainly because singers begin their training so much later; it is for this reason that singers merit special consideration.

The earliest of these reports was concerned specifically with the training of opera singers and was by Lord Bridges' Committee, commissioned by the Arts Council and published in 1964; this report led to the setting up of the London Opera Centre (Bridges 1964). The second report on that subject, the Willatt Report of 1976, was also commissioned by the Arts Council and led to the replacement of the London Opera Centre by the National Opera Studio (Willatt 1976). These reports were concerned with specialized advanced training. Two reports commissioned by the Gulbenkian Foundation dealt with wider issues, one by Gilmour Jenkins (1965), *Making Musicians,* and its successor, Vaizey's *Training Musicians* (Vaizey 1978). Among their other concerns were two of particular relevance to singers: one was that the singer's later start made LEA grants difficult to obtain, the more so as singers may not have A levels; the second was that singers, not being ready to embark on a full professional career until their mid- to late twenties, need flexible LEA grants as well as provision for in-service training opportunities with opera companies see Bridges 1965: 35–6 and Vaizey 1978: 81–2). More generally, both reports were also concerned with the fact that the music colleges were training too many students, including singers (and this point was reinforced for opera singers by the Willatt Report), that they needed high staff–student ratios and also better payment for staff. It is worth pointing out that there was the underlying assumption in all these reports that singers in music colleges were being trained for eventual work in opera.

Vaizey's Gulbenkian report dealt with the whole range of music training from primary school to advanced professional level; at the higher education level, it noted the post-Robbins development of music courses in universities, polytechnics, and other institutions and took the position that music colleges should concentrate on the training of intending performers and specialist music teachers, leaving the universities to offer more general higher education in music, while the polytechnics and Colleges of Higher Education trained schoolteachers of music; it recommended that interchange between universities and music colleges be encouraged where geographically feasible. That point was endorsed by a further study, this time on provision for all the arts in higher education, which was undertaken for the Society for Research into Higher Education (Robinson 1982); this study's recommendations for music were that the different institutions should maintain diversity rather than seek to imitate each other, with liberal non-vocational higher education being best offered in universities and more vocationally oriented courses in the other institutions. It also recommended that research was required to ascertain patterns of provision and that 'an advisory/research council be set up to gather and promulgate information on the basis of which sensible policy decisions can be taken' (Robinson 1982: 141).

Some of the recommendations of these reports have been implemented and some have not. In 1990 there was still concern about whether there were too many places for singers (and other musicians) in music colleges. Singers still had some difficulty in obtaining grants for postgraduate study. There was still no body charged with making 'sensible policy decisions' about the overall provision for the training of musicians. There has been no study that looked at both the training of musicians and the labour-market they enter. And there has been no study of the overall provision of training specifically for singers. In the remainder of this chapter provision for formal training of singers in Britain in the late 1980s is investigated in detail. The discussion of informal or on-the-job training is deferred to Chapter 5. By formal training is meant any training that takes place away from the workplace, i.e. in an educational institution or through private lessons.

The Provision for Formal Training of Singers in Britain

It is convenient to look at the provision for formal training of singers, as does the Vaizey Report, by chronological stages; the reader may find it useful to refer to that report for its discussion of provision for general musical training, bearing in mind that it was written in 1978, i.e. ten years before this study was started.

Schools

Primary and secondary schools provide some experience of choral singing, though the Vaizey Report noted a decline in disciplined class singing even in 1978; only a few schools offer systematic vocal training. These are mainly the choir schools but Chetham's School of Music, and Wells Cathedral School, both specialist music schools, list voice among their specialisms and singing is also taught at the Arts Education Trust School. In 1990 there were thirty-seven schools that were members of the Choir Schools Association. St Paul's and Westminster Abbey choir schools educated choristers only and there were more than 800 choristers in the remaining thirty-five schools which educated other boys and girls besides choristers. Child choristers, who were still mostly boys—though one cathedral had girls in the choir and another was considering having a girls' choir as well to interchange with the boys'—have to be regarded in some respects as professional singers; they earned on average half the cost of their school fees and two-thirds of them won music scholarships at public schools. They work to professional standards, learning their craft in twice daily services, the kind of work experience that it takes other singers years to acquire in adulthood, and they may earn money from tours and recording and TV work. It is hard to say how useful the vocal training that choristers get is for their adult voices; the English Cathedral Sound is highly prized in certain contexts such as Baroque and Early music but it is inappropriate for opera or Romantic music. It seems that only a small percentage of child choristers become professional adult singers and for the ones who do their vocal technique usually needs reworking, but their work experience is of lasting value.

In general, it is not regarded as desirable for school-age children to receive systematic vocal tuition, though highly desirable for them to acquire basic musical skills. However, there is considerable interest in the arts on the part of the present generation of young people and this is strong in singing too, and many courses have evolved to match this interest. Some of the music colleges' junior departments offered singing in 1990 and a special junior singing course had been set up on Saturdays at the Guildhall School of Music and Drama for 15–18-year-olds. Vocal tuition was also offered in some Local Authority Music Centres. An interesting trend in further education was the development of Foundation courses in music and the B.Tec. qualification (monitored by the Business and Technology Education Council), which were mostly taught in Colleges of Further Education and often included tuition in singing.

Colleges of Further Education

As part of the research for this book a postal questionnaire survey was conducted in the academic year 1988/9 of all Colleges of Further Education offering courses in music. The purpose was to discover the extent of provision for singing in those courses in publicly funded colleges and to estimate the potential demand for places in singing in higher education from students studying in Colleges of Further Education. Part of the motive for this was the recognition that, in order to be prepared for the quite demanding auditions students must do when applying for a place at a music college, they must have had some singing lessons and help in preparing the audition pieces. Many students have to have private lessons to prepare them for auditions to music colleges; I wanted to know the extent of provision in the public sector.

Although this section largely refers to Colleges of Further Education, a number of Local Authority-supported Music Centres were also included in this survey and, though they had a slightly higher proportion of students under 16 years of age, in other respects their characteristics were much the same as those of the Colleges of Further Education.

There were eighty-nine Colleges of Further Education listed as providing music courses in the 1988/9 *British Music Education Yearbook* (Barton and Stewart 1988); nine of these also provided

THE PROVISION FOR TRAINING SINGERS 31

higher education music courses and they are included in the survey of provision for singing in higher education discussed in the next section. I sent a simple preliminary questionnaire to all the colleges asking if they provided courses in 1988/9 that included singing. I sent a further, more detailed, questionnaire to those that responded saying they did. Sixty-three per cent of all the colleges responded to the preliminary questionnaire, of which 80 per cent said they did offer courses including singing; 42 per cent of these colleges completed the detailed questionnaire.

The colleges offered a variety of music courses, part time and full time. Most offered courses which led to examinations; over three-quarters offered A-level and GCSE courses in music and prepared students for Associated Board examinations. Half the colleges listed in the British Music Education Yearbook (Barton and Stewart 1988) offered pre-professional Foundation courses that included music and/or singing. Nearly half the colleges in the survey offered courses leading to a diploma, which could have been their own internal diploma or an external one, such as those set by the music colleges. A quarter of the colleges offered a B.Tec. in Performing Arts, and since this was a new development at the time of the survey, more may be expected to do so in the future. Some colleges offered courses which were purely recreational, a few being in pop singing; they also organized choirs and one ran a youth opera group for under-18s.

The number of students on these courses ranged from 2 to 105, with an average of 23 over all colleges, three-quarters being female. Two-thirds of all the students were in the age-range 16–24. Extrapolating from these proportions to the total for all colleges, I estimate that there were around 1,500 students doing courses which included singing in 1988/9, of whom about 1,000 were in the 16–24 age-group.

Most of the responding colleges said they were teaching vocal technique to beginners and over half claimed to offer lessons in advanced vocal technique. Half offered classes in singers' repertoire and one-third had classes in opera. As far as individual singing lessons were concerned, all the colleges offered individual lessons lasting either half an hour, three-quarters of an hour, or one hour per week. Several colleges offered additional classes in singing for two to four hours. A quarter of all responding colleges had one or two full-time singing-teachers who taught singing as part of their duties and three-quarters of these colleges employed one or more part-time singing-teachers.

Several colleges had a system akin to that of higher education music colleges, offering a one-hour individual lesson to first-study singers and a half-hour lesson to second-study singers.

It is very difficult to say what proportion of students studying in Colleges of Further Education were doing so in order to make a career as professional singers. Several of the colleges reported that a few of their students had gone on to music colleges for further study and they clearly offered serious pre-professional training. Twenty colleges gave estimates of the number who hoped to make a professional career in singing and the number of students involved averaged over these colleges was two. If this figure were representative of all Colleges of Further Education offering courses which include singing, that gave an estimated 125 students in 1988/9 hoping to make a professional career in singing, 12.5 per cent of the estimated total of students doing courses which included singing in Colleges of Further Education in the 16–24 age group. Colleges of Further Education, therefore, offered an important early training ground for young singers, and one that may become increasingly important in the future.

Associated Board Examinations

Examinations of the Associated Board of the Royal Schools of Music are important in schools and Colleges of Further Education. Information supplied by the Associated Board showed a marked increase in the number of people entering (and passing) singing exams. In 1980 585 took Grade 8 Singing; by 1988 two examinations were offered, Singing and Voice (the latter for more mature voices); 429 took Singing and 324 Voice, a total of 753 and an increase of nearly 30 per cent in eight years. There has been a tremendous increase in the number passing Singing and Voice at all grades (including Grade 8). In 1980, 2,795 passed Singing and by 1988 5,619 passed Singing and Voice, almost exactly double the number. This was further evidence of the growth of interest in singing among school-age children. However, having Grade 8 in Singing was not a requirement for entrance to singing courses in music colleges and universities as it would be for instrumentalists.

Higher Education

Singers could receive formal higher education in four different types of institutions—music colleges, universities, polytechnics, and Colleges of Higher Education. In 1990 there were eight principal music colleges, Birmingham Conservatoire, the Guildhall School of Music and Drama, the Royal Academy of Music, the Royal College of Music, the Royal Northern College of Music, the Royal Scottish Academy of Music and Drama, Trinity College of Music, and the Welsh College of Music and Drama. Other colleges of music—London, Leeds, and Dartington College of Arts and Technology and Colchester Institute—specialized in music and taught singing. In addition, many universities offered degrees in which singing could be offered as an option in the final examination, as did a number of the polytechnics and Colleges of Higher Education. I surveyed all the above institutions. In most of the music colleges, a questionnaire was followed by an in-depth interview with the Head of Singing. The universities, polytechnics, and Colleges of Higher Education were all surveyed by a postal questionnaire. This was sent to every department of music listed in the 1988/9 *British Music Education Yearbook* (Barton and Stewart 1988), thirty-four in universities (counting the colleges of the Universities of London and Wales as separate entities), nine in polytechnics, and twenty in Colleges of Higher Education.

Again, the enquiry was conducted in two stages during the academic year 1988/9; the first stage simply asked the head of music if lessons or courses in singing were taught. The response rate at this stage was exceptionally high; 94 per cent for universities, 78 per cent for polytechnics, and 80 per cent for Colleges of Higher Education. Of the thirty-two responding university music departments, only two did not make provision for singers; all the responding polytechnics and Colleges of Higher Education provided for singers. Since response rates were so high, this gave a very good idea of the overall picture about which the second stage of the enquiry, the full questionnaire, sought details. The response to this questionnaire was also good, though it was not universally appropriate and hence some departments found it difficult to fill in. That was partly because provision for singers was organised very differently, even in similar types of institutions; this is discussed in detail below. In part, also, it was difficult to apply one questionnaire to different departments because the academic context differed as between types of institutions. It is worthwhile going into this point.

In both Vaizey's Gulbenkian report (Vaizey 1978) and Robinson's *The Arts in Higher Education* (Robinson 1982) a distinction was made between vocational and non-vocational music courses in higher education, with the conventional view being that performers' courses were vocational and should be offered in music colleges, teacher training was vocational and should be offered in polytechnics and Colleges of Higher Education, and university courses were part of liberal higher education and hence non-vocational. This view was out of date in 1990: many university courses included a performance element and there were graduate courses in music colleges, which students with A-levels were encouraged to take regardless of whether they wish to become performers or teachers; some joint courses between universities and music colleges which cut across traditional lines of demarcation had also been set up and new courses had been developed. It was therefore difficult to make a clear distinction, even on educational grounds, between vocational and academic courses. Furthermore, there is no economic distinction between the two; from the economic point of view, all courses in higher education are formal general training. Nor does the market for singers recognize the distinction either; employers do not prefer one type of training to another, and, indeed, may not even ask about paper qualifications; what counts is how well a singer sings and performs. That depends to some extent on the number and quality of the singing lessons offered and on the rest of the package—languages, repertoire, movement, choral singing, etc.—and on the type of employment that is being offered. This apart, it is interesting to note that in universities, singing was nearly always part of a music degree, whereas in polytechnics and Colleges of Higher Education it might be part of a degree in Performing or Creative Arts, Humanities, or Education. Thus the type of education associated with singing lessons differed according to the type of institution. The length of the course also varied, with three years being the norm in university first-degree courses, and three or four years elsewhere. In the music colleges extensions of one or even two years to the undergraduate course were common, as many students moved on to postgraduate study;

in fact, it was not always easy to distinguish the two. The typical length of study reported in music colleges was four years.

It was not straightforward to get at the annual total output of singers from all these institutions because practices varied so much. It was decided to opt for asking for information on the number of students who obtained a qualification or completed courses in 1988 with solo singing or opera as a first study. Therefore postgraduates, first-degree graduates, and diplomates were all mixed in together regardless. This figure should, however, represent the output of all the institutions for 1988 and this was a start in the quest to discover the annual flow of supply of singers. However, whether they would enter the labour-market as singers is another matter and it is naïve to associate one thing with the other. To ascertain the supply of singers one needs to know (1) what proportion of formally trained singers try to enter the singing profession and (2) how many others seek to enter through other routes (private lessons etc.). Thus, the figures of the output of students from courses with singing as a main study do not represent the supply of singers in 1988. This question is discussed further in Chapter 2.

Music Colleges

The reported output of singers from the principal music colleges in 1988 was approximately 200, of whom one-third were men and two-thirds women (see Table 1 in Appendix 1 for details). It is interesting to compare this figure with statistics published in the Willatt Report (Willatt 1976); in 1975/6 there were 250 students trained in singing in these same colleges. I do not know if their measure was the same as the one adopted in this book, as the report did not give any definition of what 'students trained' means. My figures were for the number of students qualifying or completing courses in singing and this somewhat underestimates the number of singers who have received training, since some do not gain qualifications, either because they are unsuitable, dilettante, or because they were offered work before completing the course. The alternative measure, though, of the number of places on singing courses would overestimate output, since it would not take account of students dropping out from courses before gaining a qualification, something which is actually encouraged by colleges if they feel a student will not make the grade as a singer. Nevertheless, it seems likely that numbers of singers being trained in the principal music colleges fell between 1976 and 1988 and it was expected to fall further, as some colleges planned to cut places for singers.

Universities, Polytechnics, and Colleges of Higher Education

For the output of students who took a course that included singing as a final option in the universities, polytechnics and Colleges of Higher Education, we have to rely on estimates based upon returned questionnaires. Details are provided in Table 2 in Appendix 1. Between them the universities, polytechnics, and Colleges of Higher Education produced an estimated 143 graduates in 1988 who obtained a qualification in which singing was offered as a first or main study. This is a startling figure in view of the output of singers from music colleges. It seems likely that no one has really taken account of the fact that many institutions acting individually and accommodating only a few singers can together produce this total figure. Many of the institutions approached in fact replied that the questionnaire did not really apply to them since they had so few singers. Perhaps they will be surprised by the cumulative effect.

Interestingly, there were larger numbers of second-study singers reported in the universities, polytechnics, and Colleges of Higher Education than in the music colleges. It is hard to estimate what the total of second-study singers would be scaled up to the whole of the higher education sector; perhaps it is not relevant to do so, since it would no doubt be argued that second-study singers do not intend to make a career of singing. That may be so, but many singers have actually entered the profession through starting singing as a second study.

There were wide variations in the number of singing students as between individual institutions of higher education; while many departments had only one or two singers, others had many more. There were greater year-by-year fluctuations of the number of singing students in universities, polytechnics, and Colleges of Higher Education than in music colleges, mainly because the music colleges could easily fill their places, since demand for places was high and places were earmarked for singers. But there were also year-by-year differences in the number of first-study singers at music colleges. In 1989 there was no clearing-house arrangement for music colleges, and applicants could apply for a place in any number of colleges; there was usually an audition fee of around £25 charged by the college, but that was only a slight deterrent and did not prevent multiple applications (a fact which any figure on the number of applications per place must take into account). This partly explains fluctuations in student numbers, since music colleges offering places did not have much basis for knowing which college an applicant would choose finally (though they did exchange information on this, despite being in competition for the best students). But some music colleges preferred not to fill places if they did not deem enough applicants to be suitable; in a poor year, therefore, numbers might be down.

Entry requirements varied as between the different types of institutions. The music colleges all required an audition, and this was also mentioned by a majority of the responding polytechnics and Colleges of Higher Education. The responding universities all required A-level music and a few respondents in each sector also required a pass in Grade 8 Associated Board.

The Content of Singing Courses in Higher Education

Having got into an institution of higher education, be it music college, university, polytechnic, or college of higher education, how different was the educational and training experience? A number of questions were included in the questionnaire to all these institutions in an attempt to gain information on this point, and in addition I got a detailed account of the curriculum of the music colleges visited. The questions related to the hours of lessons, both individual and group lessons, devoted to singing. Taking individual singing lessons first, the music colleges all offered at least a one-hour lesson to first-study undergraduates (degree and diploma students). In the universities and polytechnics half the departments (in each sector) offered one-hour lessons to first-study students, with the other half offering anything from half an hour on average up to one hour; 40 per cent of the Colleges of Higher Education had one-hour lessons, with 50 per cent offering three-quarters of an hour, and the remaining 10 per cent of the Colleges of Higher Education offered half an hour. When comparing how much tuition singing students received in the different institutions, account also had to be taken of the number of weeks in the year during which teaching effectively takes place; this could be as low as twenty-five in universities, whereas it was thirty-five or thirty-six weeks in the music colleges.[2] As far as second-study singers were concerned, the standard provision was half-an-hour individual lessons, though it was more in one or two places.

However, to say that singing lessons were offered is not to imply that they were provided by the department itself. In music colleges all singing lessons took place on the premises, as it were, and all the principal music colleges had at least one full-time singing-teacher; (there was, however, no full-time singing-teacher in one smaller music college that I visited).[3] The universities were another matter; none had a full-time singing-teacher and only half had part-time teachers either, preferring to send their students to private lessons outside. In those departments where there were part-time singing-teachers they taught an average of four hours. However, in six universities there were arrangements for students to study singing in the neighbouring music college. A question was also included about how much part-time teachers were paid; in the music colleges in 1989 it was between £12 and £18 an hour. The university average was £12, and £14 the average in both the polytechnics and Colleges of Higher Education.

Turning now to the number of hours a week students spent singing either with a coach or in group lessons (song classes and the like), this varied a lot as between the different types of institution. The information on the questionnaires returned by the universities turned out to be unclear, except in one case, an Oxford college, which is interesting to cite; there, students (choral scholars) were working for nine hours a week in the college choir. In fact all the universities required singing students to do choral singing. In the polytechnics, half reported having zero hours of group lessons involving singing and the remaining half had two to three hours of group lessons; most did not require students to sing in a choir.

Similarly, only half the Colleges of Higher Education offered group lessons and the like and they took up one to two hours a

---

[2] The length of terms in music colleges depended on which particular body funded them; see Ch. 2.

[3] The number of full-time singing-teachers varied from one to five, the remainder of teaching being undertaken by part-time and visiting teachers, who averaged ten hours' teaching each. Half the responding Colleges of Higher Education had one or more full-time singing-teachers, with a range of one to seven part-timers doing an average of eight hours each; in polytechnics, two reported having full-time singing-teachers with an average of one part-timer teaching six hours.

week; but, all of them required students to sing in a choir. Choral singing was, however, more controversial in the music colleges and it is in relation to this that one fundamental difference between music colleges and the other institutions of higher education which included singing in their courses emerged: the music colleges aimed to produce soloists whereas the others did not. Most singing-teachers in music colleges were concerned to develop the solo voice and many were opposed to their pupils singing in choirs. This was no doubt one reason why professional choirs often found students trained in music colleges unsuitable; conductors and fixers reported that they could not blend vocally nor did they have the experience of sight-singing that frequent choral work brings. This was one reason why ex-choral scholars were so favoured by chorus-masters and conductors. Yet this was, and probably still is, the major source of employment for singers. One or two music colleges had begun to offer specific opportunities for small ensemble or choir training. On the other hand, students in music colleges did often get experience of singing in an opera chorus. Practice varied with respect to the teaching of opera by music colleges; in some, opera was taught on a specialist (possibly postgraduate) course and at the Royal Northern College of Music all singing students studied opera throughout their course. But even if a student was not specifically studying opera, either because the voice was not yet ready or because the course did not cater for it, she or he would usually have the opportunity to sing in the chorus of several operas during her or his time in college in a professional-level opera production.

Students in music colleges had high class-contact hours compared to their counterparts elsewhere in higher education, usually in excess of fifteen hours; singing in the opera chorus might well be in addition to that. (Also, students would be expected to practise on their own for about two hours a day.) It is hard to say, therefore, how many hours of singing students did in addition to their individual lessons. It has to be said, in parentheses, that the answer for some was very few: all the heads of singing reported that they had difficulty getting students to attend classes and to perceive the importance of studying anything other than 'the voice'. It is perhaps easier to tackle the question from another angle and say how many hours of academic study singers were usually expected to do in music colleges. This of course varied somewhat depending upon which year of the course was being considered and according to the type of course; graduate courses tended to have more academic input than performance courses; a postgraduate opera course would have much less. In most music colleges, students did about six hours a week of academic study. This consisted of languages—French, German, and Italian—musicianship, and the history of music. Other classes were in speech, movement, and stagecraft; some included Alexander technique; all had repertoire and song classes. Students on graduate courses, which were designed for training teachers, did more hours of musicianship (aural training, sight and score reading, transposition, etc.) and, at least in two major colleges, students had to learn how to teach singing and about physiology of the voice.

In addition, there were classes in opera; in those music colleges in which opera was taught as an integral part of the undergraduate courses, opera could take up to four hours a week; on postgraduate courses, more time again, being combined with coaching sessions for students to study roles.

Let me now summarize what has been said about the provision of places for singers in institutions of higher education. In 1988/9 in Britain there was a spectrum of possibilities for a student wanting to study a course that included formal training in singing. This spectrum moved from the more vocational through to the more academic. It is usually assumed that the performance courses in music colleges are the most vocationally oriented and the university courses are the most academic, with the polytechnics and Colleges of Higher Education lying somewhere between the two. But for the concept of vocational training to be meaningful it is necessary to take account of the different types of employment on offer for singers and in fact they cut across this supposed vocational–academic spectrum. Say, for the sake of argument, that work in the singing profession divides up into opera, concert and recital solo work, choirs and sessions work, and teaching: an opera soloist is very unlikely to sing in a Baroque choir, a teacher to sing in an opera chorus, and so on. And the skills needed for the different specialisms tend to become specific to the type of work with on-the-job training and experience. To sing in an opera chorus you need to memorize the words and music and to cope with stage moves, etc. at the same time; to sing in a choir you need to be an excellent sight-reader; to do a solo recital you need a good sized repertoire, an excellent memory, and the stamina to sing with little break for up to an hour. Teaching has its own set of demands again. The skills needed for all these vocations could be acquired in a different mix in any of the institutions of higher education discussed above. It is therefore not useful to make the distinction between vocational and academic provision for training for singers. This is discussed further in Chapter 2.

There is one educational aspect of the different types of provision for training singers that should not be ignored and that is that the quality of the educational experience of one student is affected by the abilities of the others. There is surely a critical number below which student numbers should not fall. All heads of singing spoke of the danger to students of being a big frog in a little pond and this problem gets worse the smaller the number of students. The labour-market for singers is large, demanding, and cruel and anyone who is not prepared for that will not only fall by the wayside but will feel cheated and betrayed. This problem is the worse for singers studying with only one or two other singers in their year and perhaps not more than four or five in the department. Numbers of singing students in many of the universities, polytechnics, and Colleges of Higher Education were very low. Taking ten students a year completing courses which included singing in 1988 as an arbitrary cut-off point, two of the music colleges had less than that number, but only three of the other institutions (one university, one polytechnic and one College of Higher Education) had as many as ten or more students offering singing as part of a final qualification in 1988. Although this way of looking at things can be very misleading—for example, one well-established College of Higher Education, with its own school of music, had seven first-study singers but twenty second-study singers in 1988—it is a point that needs considering, from the educational side as well as in relation to the costs of training.

Advanced Training

It is hard to say where postgraduate study ends and advanced training begins; training that students, usually postgraduates but also undergraduates, get in music colleges can be sufficient to enable some to enter full-time professional work in singing on leaving. The two-year opera postgraduate opera courses run by several of the major music colleges would give students the experience of taking part in a number of full length opera productions with all the preparation of vocal and language coaching, production rehearsals, costumes, and make-up that are entailed. By the time they leave postgraduate study many students will have spent six years in college. Is that not enough? Here we come back again to the effect of the late start that singers inevitably make; a violinist leaving college aged 22 has studied for well over ten years all told; by the age of 25 a singer has had only half that; that may be enough for some singers but not for all and this depends on the type of work to be undertaken—training for opera takes longer and it takes longer to train a soloist than a choral singer. In one sense training never ceases, as an artist will go on learning from experience throughout his or her life. Informal, on-the-job advanced training may last for any length of time. Be that as it may, there is specialized provision in Britain for formal advanced training beyond postgraduate work in music colleges.

The National Opera Studio

The National Opera Studio was started in 1978 as the successor to the London Opera Centre which used to take around forty students on a two-year course; the Studio offers a ten-month course to twelve advanced students each year. These students may be either postgraduates who are on the brink of a professional career or young professional singers working for, or under consideration by, opera companies. The Board of Management consists of the directors of the six main British opera companies, each of which sponsors a student annually. The Studio is subsidized by the Arts Council; in 1990/1 its grant was £100,000. It also had funds from the opera companies, from trusts, and from sponsorship; fees were £4,000 in 1989 and students usually had their fees paid out of these funds one way or another. Students might get support for their living expenses from these sources, or they relied on earnings if they had an engagement or on their own or parental resources. Entrants averaged 27 to 28 years old. They had usually been to college for four to six years and had some professional work experience, such as in the Glyndebourne Festival Chorus doing small parts or covering roles (understudying). Applications had to be accompanied by two recommendations from opera companies, senior staff in music colleges, singing-teachers, conductors, or other well-known professional musicians and administrators and there were two rounds of auditions. In the first round, about 120 applicants were heard, and

35 in the final round; though selection was made on the basis of the best voices and presentation, nevertheless the Studio tried to have a balance of voice types. In 1989/90 there were 4 sopranos, 2 mezzos, 3 tenors, 2 baritones, and 1 bass; in 1988/9 there were 6 sopranos, 3 mezzos, 1 tenor, 2 baritones, and 1 bass. These were selected from a typical audition list of 50–60 sopranos, 20–30 mezzos, 15 baritones, 10 tenors, and 3 or 4 basses. These figures are quoted as a demonstration of how much more competition there is for women, partly because there are always more of them (especially sopranos) and partly because there is proportionately less work.

The Studio trained individuals in suitable specific roles and in scenes of operas. To some extent, therefore, it avoided the problem of the balance of voices inherent in the production of one whole opera. This enabled the Studio to tailor workshops and the showcase scenes to the students' best advantage. Students received intensive coaching in the roles selected for them to work on besides language tuition, movement and stagecraft, and general musicianship. No vocal tuition was given as students were expected to continue with their regular teachers.

Conceptually, the Studio to some extent falls between two stools; it is neither higher education nor part of the performing arts sector. In economic terms it offers general training but nevertheless training that is specifically vocational. The Willatt Report (Willatt 1976) had envisaged training links between the Studio and opera companies in the form of apprenticeships for singers with opera companies. Many people in the singing profession would still like to see such links develop. As the music critic Hilary Finch wrote in an article in *The Times* (14 March 90) 'there is growing concern about the isolation of all training, particularly of the National Opera Studio itself, from involvement in the day to day work of the opera houses'. This concern has been voiced by singing-teachers, opera company administrators, and by singing students themselves. However, the opera companies have neither the economic incentive nor the financial resources to do training of this type; but this is where the application of the economic analysis of training comes into play; it shows that no one opera company has an incentive to pay for general training and by using the facilities of the National Opera Studio they can enjoy the benefits of having trained opera singers without having to pay for the full cost of the training. This is discussed in detail in Chapter 2.

In other countries there are opera schools attached to opera-houses. A report by Nieuwenhuis in 1990 for the Ministry of Culture in the Netherlands provides a thorough survey (in Dutch) of opera studios in Europe (Nieuwenhuis 1990). They ranged from the German model, with the Opera Studio being run by the main opera-house with students financed by bursaries and fees for work in main house productions, to summer schools. The advantages of the German model are that students gain on-the-job work experience in big professional companies; the disadvantages are that students may be overworked or possibly even exploited and the training offered is geared to the needs of the opera-house rather than those of the students (see Appendix 3). They do nevertheless avoid the hot-house, unreal atmosphere of the British National Opera Studio model, in which students are essentially cut off from the day-to-day work pressures of the opera-house. The question of how best to organize advanced opera training probably may never really be solved. It is certainly one that seems to crop up every fifteen years or so; reading the Bridges and Willatt Reports (Bridges 1964; Willatt 1976) in the 1990s gives one a strong sense of *déjà vu*, as do the criticisms voiced by Joan Cross and Ann Wood thirty years ago, to which the Bridges Report was a response.

British Youth Opera

British Youth Opera (BYO) was founded in 1987 as an advanced level summer course. In 1989 the course lasted six weeks in which two operas were studied, after which, British Youth Opera toured for three weeks, giving thirteen performances. It catered for singers and instrumentalists between 22 and 30 years old and managed not to charge them for their training by a feat of financial juggling of box office, sponsorship, and grants of various kinds. Of £450,000 raised between 1987 and 1990, only £18,000 was from public funds. In 1989 twenty-two singers took principal roles with a further ten doing understudy and ensemble work; they were selected from approximately one hundred applicants, practically all of whom had studied at a music college or conservatory. British Youth Opera has attracted a considerable amount of interest both inside and outside the singing profession and was expanding its activities at the time of writing.

Britten–Pears School of Advanced Musical Studies

This school was founded in 1972 by Benjamin Britten and Peter Pears and has had a strong tradition in singing. Courses for singers, twelve at a time, were for advanced students and young professionals and took the form of eight or nine days of intensive masterclasses with distinguished guest teachers on various types of song, oratorio, and opera roles. Each year one opera was studied and performed; the opera course lasted for just over three weeks. There were ten courses in all between April and October in 1990; tuition fees were £200 (£320 including accommodation, for the opera course). In 1988 sixty-two singers participated in courses, half being British; two-thirds were women. The typical age-group was 22–32 years old, and the students had almost all studied in music colleges and several had already had some professional work. Auditions were held throughout Britain and also in the USA and Canada; in 1988 there were around 200 applicants. A bursary fund enabled a few students to receive help with payment of fees.

Other Advanced Formal Training

In 1990 there was a variety of other arrangements for post-college formal training; in London, Institutes of Adult Education ran courses in opera and singing which, although primarily designed for amateurs or semi-professional adults, were often used as a training ground by music college or other such graduates waiting to get regular professional work; examples were Morley College Opera (which took forty a year for six hours' tuition a week, putting on three performances and two workshop performances a year; 90 per cent of participants had been to music college or received equivalent training; there were at least three applicants for each place); Abbey Opera was in cold storage, but in the mid-1980s it had fifty to sixty regular singers, mostly aspiring professionals and singers already doing some professional work doing several productions a year, including at the Camden Festival. The Mayer–Lismann Opera Workshop (twelve to fifteen students, six hours a week, preparing scenes from operas and giving workshop performances); Opera Viva through which many now well-known singers passed in its heyday; Beaufort Opera; Opera Integra ... there were many others of very variable standard. These all charged Adult Education course rates, which were very reasonable for local participants, though much more expensive for those residing or working in other Boroughs.

There were also many professional-level summer-school opera and song courses—Dartington Summer Music School, Music at Oxenfoord, Summer Music (*Classical Music* lists over fifty in its annual Guide to Summer Schools). Rather special was Clonter Opera Farm Opera Studio; founded in 1974 it gave high level intensive training in scenes from operas, followed by public performances (in the former silage shed) for twelve singers who were just starting a professional career. A Music Trust enabled all those chosen to participate free of charge; students were auditioned in London, Glasgow, and Manchester. Besides all the British offerings, there were international summer school courses throughout Europe and the USA, which young British singers may attend; some gave financial help to participants (details are given in the *British Music Yearbook* (Carter 1990)).

There was, therefore, a range of possibilities for advanced training; what was on offer was usually specialized and taught by specialists. Some courses were subsidized out of public funds and some not, but often students, even on subsidized courses, had to pay their own fees and other costs.

Private Singing-teachers

Of all the arts, the art of singing is perhaps one of the least understood and consequently the art of teaching singing is particularly controversial. As this book is concerned only with the economic aspects of the singing profession, the focus here is on how the labour-market for singing-teachers works. Being an entirely free market, it in fact works according to the principles of supply and demand with fees for private lessons being determined by those forces. Whether the fee represents value for money is not something that can be discussed without getting into the difficult area of the quality of singing-teachers. It is very much a case of 'you pays your money and you takes your choice', though it is very hard to make an informed choice and much damage can be done by the wrong one.

Private singing-teachers cater for a very wide spectrum of pupils indeed; this ranges from beginners to advanced professionals, and

THE PROVISION FOR TRAINING SINGERS 47

includes young people hoping to go to college to study singing, college graduates who are studying privately while doing professional auditions, and established professionals (most of whom have regular singing lessons throughout their careers), as well as amateurs. Some singing-teachers specialize in one end of the market or the other; some teach a mixture of age groups and abilities. In 1990 there was no bar to becoming a singing-teacher; those who were accepted on the roster of the Incorporated Society of Musicians had received professional musical training but that did not necessarily mean that they had learned how to teach, nor how to teach singing. Courses in two music colleges, the Guildhall School of Music and Drama and Trinity College, specifically taught their graduate-course singing students to teach singing but that was a recent development and the effects had not yet filtered through to the main body of singing-teachers. As part of the research for this book, a postal questionnaire survey was undertaken in 1989 of members of AOTOS. Not all singing-teachers in Britain were members of AOTOS, though it was the main professional association specifically for singing-teachers. Nevertheless, AOTOS drew its membership from a wide range of singing-teachers both regionally and in terms of the type of work they did (teaching in music colleges, other institutions of higher, further, and adult education as well as exclusively private teaching). It was an active association and there was every reason to suppose that its members were representative of the profession as a whole.

A short questionnaire was sent by AOTOS to each of its 300 members. There was a 32 per cent response rate, which is regarded as reasonable for this type of survey.

Besides working as singing-teachers, 60 per cent of respondents were active as performers; they spent just over half their time in teaching singing which they did for a reported average of twenty hours a week. In addition, 31 per cent of all respondents taught something other than singing, such as piano or aural training. As to where they taught, a third of all responding singing-teachers taught in institutions of higher education, i.e. in a music college, university, polytechnic, or College of Higher Education, teaching an average of twelve hours a week. Just under a third taught in a Further Education College or school, nine hours a week on average. All but four respondents taught privately, working an average of twelve and a half hours a week.

Many teachers, of course, taught in more than one type of institution, as well as having a private teaching practice. These average figures to some extent mask the considerable variation that there was between individuals, with the lowest reported hours being two and the highest fifty-two a week for all types of work. The breakdown of total hours worked teaching singing is given in Table 3 in Appendix 1. They show that over half the singing-teachers surveyed taught between ten and thirty hours a week in 1989.

The number of private pupils that teachers had varied very considerably according to how active teachers were, the proportion of their time spent on teaching privately and also on how often the pupils came for lessons. The average number of pupils was twenty-three per teacher, of whom a third were male, two-thirds female. (This ratio of males to females, incidentally, was exactly the same as that for singing courses in higher education.) If the survey of respondents was typical of all AOTOS members, this suggests that there were nearly 7,000 people taking private singing lessons, plus an unknown number studying with non-member singing-teachers. Since teachers reported that on average a quarter of their pupils hoped to make a career in singing, this adds up to a large number of aspiring singers!

In order to get a rough idea as to how many young people were having singing lessons, possibly with a view to entering the singing profession, teachers were asked to indicate the age groups to which their pupils belonged. The results are presented in Table 4 in Appendix 1, which shows that in general female pupils tended to be somewhat younger, but this is not very marked. Although it is impossible to say from these data how many private students will become professionals, it is nevertheless likely that many do try to get into singing by studying privately. (It should be said here that the questionnaire did not ask about pupils under 16 years of age and so they have not been taken into account in these results.)

Singing-teachers were asked for their recommendations as to the number and length of lessons. Nearly all recommended lessons lasting half an hour for beginners, and were more or less equally divided as to whether there should be one or two lessons a week. For more advanced pupils, 56 per cent of respondents recommended one lesson lasting one hour; 20 per cent recommended two lessons lasting one hour and the remainder recommended anything from one half-hour lesson to three one-hour lessons a week.

Turning now to the qualifications and experience of singing-teachers; 80 per cent of respondents had a formal qualification as a teacher and 67 per cent had a formal qualification as a performer; 50 per cent had both; 10 per cent had no formal qualification as either. As to whether they received training in teaching technique on the course leading to their qualification, this was more difficult to interpret. The question was, in fact, ambiguous in that it did not specifically ask for training in the technique of teaching singing. People who had a general teaching qualification and had done some teaching practice may have answered yes on that basis. Therefore, it seems likely that the answers to this question gave a rather rosier view than the true position merits. For what it is worth, 53 per cent of respondents said they had had training in teaching technique. The other question was about years of teaching experience; the average of these was eighteen years, but with answers varying from three to forty-five years the average does not signify much! Breaking this down somewhat, 57 per cent of respondents had taught for ten to twenty years, 22 per cent had taught for under ten years, and 21 per cent had taught for twenty-one years or more.

Besides individual private teachers, there are also a number of private academies of singing; those listed in the 1988/9 *British Music Education Yearbook* (Barton and Stewart 1988) were surveyed by postal questionnaire. Though the response rate to this was good, the organizations were so varied that it is impossible to summarize the findings; they ranged from the highly professional to the virtually charlatan.

A few of them had quite significant numbers of pupils; the biggest had 140 students in singing of all ages, 50 per cent being between 16 and 24 years old. This school employed five part-time singing-teachers as well as distinguished visitors who ran masterclasses from time to time; 30 per cent of their students hoped to make a professional career. Another offered postgraduate and professional training; it had over 100 students, 80 per cent of whom were between 25 and 35 years of age, and 90 per cent of whom hoped to make a professional career; it employed two full-time and six part-time singing-teachers and students attended for four hours a week. Another had nearly 100 pupils of all ages and employed one full-time and three part-time singing-teachers; 10–20 per cent of its pupils hoped to make a career in singing. Fees in all these academies/schools, etc. ranged from £10 to £30 per hour in 1989.

The private market in singing lessons is important to this study for two reasons: one is that singing-teachers form part of the singing profession; their earnings represent an alternative or additional source of income to performance and this becomes increasingly important as singers get older; the second is that the cost of singing lessons and coaching are an ongoing professional expense for singers. These points are developed further in subsequent chapters of the book.

Conclusion

This chapter has given a detailed account of the provision of formal training for singers in Britain at the end of the 1980s. Nothing has been said so far about the opportunities for informal, on-the-job training but this is best discussed in the general analysis of the demand for singers in Chapter 3.

What this survey shows, especially at the level of higher education, is that the diversity that was noted as flourishing by Robinson (1982) has now begun to go to seed and it would seem that some pruning is needed. The music colleges have responded to earlier criticisms of unfettered growth and in singing, at least, numbers being trained have fallen over the last 15 years; at the advanced level, the National Opera Studio trains fewer students than did its predecessor, the London Opera Centre. But there is still the incentive for the growth of courses which include singing in the universities, polytechnics, and Colleges of Higher Education. There is a considerable excess demand for places to study singing, and since the higher education system is demand led and, furthermore, institutions are being encouraged to respond to demand, there is every possibility of them expanding courses in which singing is an option. This situation is apparently not being monitored and the cumulative effect of uncoordinated increases in student numbers can, as I have shown, be surprising; in 1988 40 per cent of all students doing formal higher education courses with singing as a first study or option were outside the music colleges. This proportion is very likely to increase if music colleges cut down their numbers and the other institutions increase theirs.

Vaizey (1978) took the position that only students trained on performance courses could be regarded as entrants to the music profession and more or less implied that such courses only existed in music colleges. That position does not take account of the way that the market for singers works. For the development of performance and other such courses outside the music colleges not only shows that there is considerable demand for them but also that students find them useful. How can courses be evaluated for the economic benefits they offer? In other branches of higher education the usefulness of courses, at least in terms of helping students to get work on leaving college, can be measured by data on the 'First Destinations' of students six months after graduating; statistics on the number of recent graduates finding work and the type of work are collected by all universities and polytechnics and listed by the type of degree. Such data are collected for music and for performing arts but not specifically for singing. However, these data are quite inappropriate in a labour-market in which there is little regular work. (The difficulties of measuring employment in singing are discussed in detail in Chapter 3.) In the past, music colleges have not all been required to collect this type of information but most were beginning to do so in 1990; however, it is not an easy business to follow up former students. The music colleges and the National Opera Studio keep records of the engagements of their trainees but these never really give a full picture of the record of all students, only of the successes. The Studio claims a high rate of success with placing students but does not define how that is measured. Listing students' engagements does not tell you how many performances they are doing or if they are working often enough to earn a living. Thus, 'First Destinations' data are of no use in estimating the flow of supply of new entrants to the singing profession.

It seems inevitable that music colleges and other such institutions see themselves as successful if they produce stars; after all the performing arts are very much geared to the star system. But it is saddening to see the star system at work in the early years of training. Music colleges perceive the need to attract good students in order to produce good graduates and diplomates and this is one of the purposes of their showcase opera productions which very much embody the star system. It is, however, hard for students who do not flourish early because they often feel cast by the wayside; the result is that there are many now successful singers who have angry memories of their college days. Another form of window dressing that goes on in music colleges consists of listing in their prospectuses well-known teachers and singers giving masterclasses without indicating the extremely part-time nature of their commitment. These are indications that, although music colleges are increasingly being integrated into the higher education sector, they do not yet fully subscribe to the educational norms of that sector. Perhaps they should not do so nor be expected to. They are, however, in receipt of public subsidies, in fact on more generous terms than other types of institutions of higher education (see Chapter 2) and so might be expected to share the same educational objectives and achievements.

These achievements are typically perceived as relating to success in the labour-market. But unlike other labour-markets, the labour-market for singers pays little attention to what is called the 'screening function' of higher education. In most professions, a paper qualification from an institution of higher education is regarded by employers as a signal that the graduate or diplomate has certain characteristics which employers seek. According to the screening hypothesis, it is not the content of courses that is of value to employers so much as general achievement—hard work, the ability to perform set tasks, communication and social skills. Thus the screening hypothesis clearly questions the validity of the notion that vocational training is a necessary requirement for entry into a profession or occupation and denies the distinction between vocational and general training. In artistic professions paper qualifications do not seem to provide the correct signals to potential employers and this is very much in evidence in the singing profession. Employers, i.e. opera-houses, chorus-masters and fixers, conductors and suchlike who hire singers, show little interest in paper qualifications or even attendance at a music college or other institution of higher education, but prefer to rely on their own assessment of singers in auditions or at performances. This is a much more costly way of screening future entrants to the profession than reliance on the higher education system to screen entrants for them, and the explanation must be that a degree or diploma in singing is an ineffective screen for the qualities in singers sought by employers. This may be because the institutions of higher education are not sufficiently selective but it is more likely that it is inherently difficult to measure artistic quality by achievement in examinations.

This is further cause to question the equation of the output of graduates and diplomates from music colleges and other institutions of higher education with the annual flow of supply of new entrants in to the singing profession. But, though it may not be possible to correlate employment with training in the same way for singers as it is for other types of graduates and diplomates, there must nevertheless be some concern about the extent to which the training of singers leads to employment. That is why it is necessary to understand the labour-market for singers.

2
The Cost of Training Singers

Another factor that might be expected to influence the supply of singers is the cost of training. In general we would expect that the higher the cost, the lower the number who would undertake training. The fact that places in higher education are subsidized and that many singing students get a student grant for their undergraduate course undoubtedly leads more to students studying singing than if they had to pay the full cost themselves. But postgraduate students, who are arguably the most serious candidates for entry into the singing profession, are not entitled to student grants and so many singers bear the cost of advanced training themselves, unless they can get a college bursary or other private assistance.

There are in fact three aspects of the cost of training singers: (1) the direct costs of formal training, (2) the indirect costs of formal training, and (3) the costs of informal training. Only the first type of costs is subsidized by educational grants in Britain; the second type is borne entirely by the trainee singer and practically all the third type as well, though there could be some (small) element of public subsidy via individual arts organizations. A further set of costs relating to 'training' in the sense of developing one's art continue throughout a singer's career, the cost of singing lessons, coaching, etc., but these are best dealt with later on in relation to net earnings.

But before going into detail about the magnitude of these costs, the economic function of training is discussed. This does not involve complex economic analysis but is rather a particular way of looking at things; it sheds light on several important aspects of training, particularly on the question of who really bears the cost of training.

The Economic Analysis of Training as applied to Training Singers

There is usually an economic explanation of why things work in a certain way and that leads to the question whether that is the best or only way. For example, one could ask the question 'do we need music colleges to train opera singers?' Answer, no, they could be trained by opera companies or privately (as they were in the past). So why is it that music colleges run opera courses while opera companies have few trainee singers? The economic explanation is that opera companies have no incentive to pay the cost of training singers since they cannot capture all the benefits of that training; once they have trained, singers can go and work elsewhere using the skills they have learned. Opera companies will give young singers the opportunity to gain performing experience but will pay beginners less than they would pay established singers, thus passing on the cost of the learning experience to the young singer. 'Passing the buck' on training costs is widespread throughout the economy and manifests itself in many ways, but there is always a good economic reason behind it. In general, employers try to avoid training costs if they can, because training is not profitable; but even though arts organizations are not profit-making enterprises, few find it economically possible to operate their own training schemes. The National Opera Studio acts for all the main opera companies and is supported by the Arts Council and by private sponsors. Undergraduate training is provided mainly by the subsidized higher education sector or by private teachers paid directly by the trainees. Even advanced professionals have to pay a high proportion of their learning costs themselves; freelance singers are usually expected to learn repertoire in their own time and at their own expense; only singers who have a regular contract, such as those in an opera chorus, are normally paid for learning time and have the cost of coaching paid by the opera-house. This is because freelance singers are able to benefit from their investment of learning a role or whatever by performing it with different opera companies. On the other hand, they are paid for rehearsal time for a specific production because they would not otherwise have a financial incentive to attend rehearsals. There is an economic explanation of why training and learning costs are passed on to the individual or organization who is most able to benefit.

The economic theory of training makes strong predictions, which are clearly relevant to the training of singers and other musicians. A fundamental distinction must be made between 'specific' and 'general' training; specific training is that which raises a trainee's productivity only in the enterprise providing the training, while general training improves performance in any enterprise. Economics predicts that where training is specific, the costs of training will be borne by the employer, whereas the costs of general training will be passed on to the trainee, since no employer will have the incentive to finance training which can be used by the trainee in other enterprises.

This is a subtle idea and it is worth spending some time explaining it.

Training which helps the trainee to perform tasks better by teaching him or her things that are specifically useful to one particular employer is specific training. Examples of this are orientation courses in the workplace, the acquisition of skills which are useful only in one enterprise and the acquisition of information about the internal organization of a particular enterprise. For instance, the technical personnel in a theatre or opera company have to learn about the technical aspects of one particular theatre; each production of an opera or play requires specific knowledge of the technicalities of that production. On the other hand, training that improves the performance of a worker by teaching him or her skills or knowledge which apply to more than one individual enterprise is general training. In fact most training is general training and that ranges from basic education right through to the acquisition of very high order skills and knowledge. Even in the example given above, some element of learning about the technicalities of one theatre or production will brush off and enable the worker to do a job more efficiently in other theatres and productions. But the crucial point is that employers have an incentive to pay for the costs of specific training, since they can capture benefits from the resulting enhanced productivity of their workers, but they have no incentive to pay for general training. This is because an employee who has received general training can move away from the enterprise which paid the training costs and get a job elsewhere on the basis of that training. So employers will not pay for general training costs, though they may nevertheless make provision for training to take place; depending upon circumstances, the costs of general training will be borne either by the trainee in the form of lower earnings or by the taxpayer in the form of a subsidy to education.

The costs of all training are identified with a powerful fundamental notion of economic theory, that of opportunity cost. Opportunity cost is the cost in terms of earnings forgone or output lost by spending time and other resources in one pursuit which could have been expended in another. To the trainee the opportunity cost is what he or she could be earning in the time taken out for training. The cost of training to the employer is the value of the output lost while training is taking place: the supervisor is not doing his or her own work while instructing the on-the-job trainee, so output is lost: the trainee's time is not being applied to producing goods or services on off-the-job training periods, and that is true of the instructor's as well. All types of training, therefore, have a cost. The costs of training singers consist of direct costs—singing lessons, course fees for the great variety of training courses that exist both in higher education institutions and in other types of provision—and the indirect costs of both earnings forgone during the formal training period and low earnings in the early years of a singer's career. Even if training is subsidized by a student grant so that direct costs are covered, the indirect costs will still be borne by the trainee, and as will be demonstrated later, these are particularly high for singers, because they enter the labour-market relatively late.

Different types of training may be distinguished; formal training which is provided away from the workplace; on-the-job training, which is training supervised by more senior employees of recruits while they are working; off-the-job training is formal training provided by the employer for employees in work time but not while they are engaged in producing goods or services. A further category of on-the-job training is learning by doing, or learning from experience. While this is an important aspect of training, particularly for performing artists, since it happens anyway, and, as it were, cannot be avoided, the costs cannot be allocated to either employee or employer. While much of the general training that singers receive is formal and undertaken in institutions of higher education, a considerable amount of on-the-job training also takes place at the beginning of the singer's working life. This is why less experienced, younger singers tend to be paid less.

How does the distinction between vocational and academic training fit into this way of thinking? The answer is that from the economic point of view both are general training in so far as they are formal, course-based training. The fact that vocational courses are taught in colleges means that even if they are aimed at a particular type of employment, they are not geared to the exclusive specific requirements of one employer. The distinction between academic and vocational study may be educationally valid, but it is not valid in economic terms. What concerns economists is who has the incentive to pay for the training.

The basic training that all singers must have in vocal technique, musicianship, foreign languages, and a familiarity with a range of repertoire is clearly general training, since these skills would be necessary requirements of all employers of singers, be they choral conductors, opera-houses, or education authorities hiring teachers. It is interesting to note that college-trained students have a range of other job possibilities open to them by virtue of the fact that they have undertaken (any) higher education.[1] The learning habits and discipline required to study singing, even in the early stages, as well as the acquired knowledge of foreign languages and the other academic subjects studied, are all attributes valued by employers outside the music profession. In addition, singers are often very personable and energetic and find job interviews easy; again these make them an attractive proposition to employers. Thus training as acquired during a three or four year course in singing in a music college, university, polytechnic, or College of Higher Education prepares singing students with the basic skills needed for work in the singing profession, but it also is a preparation for other types of work as well. That being so, what incentive would any employer have for paying for the cost of this basic training? Even if there were no publicly subsidized training for singers, no one individual employer would be prepared to pay the cost of something from which he or she could not extract all the benefits; since there are many different employers who can utilise these skills and singers can easily move round the labour-market, not only in Britain but internationally as well, skills they acquire in one employment can easily be transferred elsewhere. The fact that the labour-market for singers is predominantly freelance reinforces this point.[2]

---

[1] The graduate labour-market responds to the general nature of much of higher education by accepting graduates of a wide range of disciplines. According to the Association of Graduate Careers Advisory Services, one third of all jobs for graduates are open to graduates of any discipline. The other side of this coin is that only some graduates of any discipline go into the field for which that discipline on the face of it prepares them; no-one expects all philosophy graduates to become philosophers! Even in more vocationally oriented subjects like engineering or science there is never a one to one correlation of degree content to occupation.

[2] This argument is borne out by the example of the National Opera Studio; no one opera company has the incentive to train opera soloists, but the industry does. Even so, it is doubtful that the opera companies would support the Studio without significant state subsidy.

This is equally true of more specialized and advanced skills. Would specializing in contemporary music, even of one composer only, be specific training in the economic sense? No, because once a singer has learned some songs and acquired the proper technique and style, she or he could perform them anywhere. The composer might want to train the singer in the art of performing his or her songs, but the composer is not usually a promoter and so does not employ a singer in the usual sense of that word. Indeed, it is very hard to think of examples of specific training in music; it would have to be something like a conductor who could only conduct one orchestra. (Even the man who trained himself to conduct just one Mahler symphony could conduct it anywhere that would have him!) For singers specializing in opera, all the stagecraft skills, the ability to work with producers, conductors, and with other singers, as well as learning roles, are all general skills which every opera company utilizes, and ones which are appropriate for musicals and other stage work too. Opera companies do groom individual singers for certain roles and sometimes pay them during their training, but that is not specific training in the economic sense, because there is nothing, except a contractual ban, to stop the singer from eventually performing the role with another company. There is a specific training element attached to a particular production of an opera and singers are required to attend rehearsals (the fee covers the rehearsal period as well as performances); this is an example of employers paying for specific training. Nevertheless, singers can also take some element of the training away with them in the form of a better understanding of the role. For all these reasons learning to sing is clearly a general training and therefore there are deep economic principles that account for the fact that singers have to bear much of the burden of the costs of their own training.

Because of the nature of artistic 'output' it is not always easy to identify the training element. Orchestras and opera companies pay for rehearsal time—is the rehearsal period part of the 'product' or training time? If it is training, how much of it is specific to the individual orchestra or opera company? Obviously, a lot of it is general since if you learn a Beethoven symphony or *Madam Butterfly*, you could perform it with another orchestra or opera company. Unfamiliar repertoire, however, may be specific to one organization and that would explain why musicians and singers, who might otherwise be expected to learn their parts on their own time, are provided with rehearsal time and coaching for it at the expense of the orchestra or opera company. (The other side of this particular coin is that well-established singers are often unwilling to learn infrequently performed pieces, since they do not expect to be able to spread the fixed cost of learning them over enough performances to make it worth their while.) It is hard to say exactly what constitutes training and what does not in each and every case but what is clear is that a large element of the training of instrumentalists and singers is general; the ability to sight read and learn roles are general skills which trained musicians are expected to possess (among many others) before they are hired. Furthermore, we can predict that as financial pressures increase, increased demands will be made on performers' skills, since rehearsal time is expensive; the sessions world already relies entirely on the high standards of general training of singers and players to perform expertly with little or no rehearsal time. It is hardly surprising that greater skills are called for, since those employing singers do not have to pay for them to acquire them; the singers are forced to bear the cost. If there were a shortage of highly trained singers, however, promoters would have the incentive to pay for more rehearsal time.

Who, then, will pay for general training? Obviously, it is in the interests of employers to have access to a highly trained workforce. It is also in employers' interests to overstate training needs, since they do not have to pay the cost of training. But it is also in the interests of individuals to undertake training on their own behalf, since by and large the more highly trained and educated people are, the more they will be paid for their work. Training and education are regarded by economists as investment in human capital; the more time that is taken out in training, the greater the skills acquired, the more productive are the services of the trained person. By specializing in a particular skill, or set of skills, workers become more efficient; the bigger the market, the more scope there is for specialization; an example of this in singing is that the world-wide growth of opera in recent years has enabled some opera singers to make a living by specializing in a few roles with which they travel round, thus cutting down on their investment of time in learning roles. The longer the training period, the higher are both the direct costs of training (in the form of fees paid to teachers, colleges, etc., and living expenses) and also the indirect costs of earnings forgone (i.e. those which could have been earned by working in the training period). The higher the costs, the greater future earnings must be to make it worth the person's while to have undertaken training. On the other hand, it is generally the case that people with more education and training are more highly paid and therefore will recoup their training costs. Economists view individuals as making choices as to how much training they will invest in, based on a rational calculation of the rate of return on their training costs over a lifetime of working. Even though this may not be true for each and every individual, it is a basis for predicting labour supply behaviour in aggregate.

Costs are the higher because many singers now are graduates and that raises the opportunity cost, the indirect cost of earnings forgone, since graduates are typically paid more than non-graduates. It may well be that this explains why fewer men enter singing courses than women; because men tend to earn more than women they have more to give up, that is their indirect costs of training are higher than for women.

There is a choice of how much and what type of training to undergo for entry to the singing profession and each route has different costs; it is clear that there are several different routes, and not all involve the formal study of singing in a music college. The route that starts in choir schools, going on to a choral scholarship, and then entering the profession directly, often into sessions work, not necessarily having had even much in the way of individual lessons (though many choral scholarships pay for individual singing lessons), is the least costly form of training. If it were not for the fact that choral scholars often move on to a music college for postgraduate training, this type of training would be virtually free of direct costs. The on-the-job training and experience that being a choral scholar offers for certain types of work is superior to formal training in music colleges, and certainly some choral conductors and fixers prefer singers who have trained by the choral scholarship route. One has to conclude, therefore, that this is a cost-effective means of training. Another relatively low-cost model is private training, where singers train with a private singing-teacher while either working outside the singing profession or doing a degree in another subject. Some people work in other professions and sing as amateurs for a while before deciding to become fully professional singers. That way they cut down the indirect costs of training.

The costs of training are determined by the length of time taken to develop vocal and other skills to the point at which it is possible for the singer to consider entering the profession and so vary considerably. Some singers have completely natural voices and talent, and the issue is simply one of discovery; it took one well-known British bass only three months of singing lessons before he was hired by Covent Garden! Others may take ten years. It is generally believed by singing-teachers that it takes four years' formal training to get a singer to the point of being ready for group work—choirs, choruses, ensembles—and six years for a soloist. But that is formal training; on-the-job training also takes place and this may be a substitute for formal training. But equally, formal training may be a substitute for on-the-job training; for example, the National Opera Studio believes that its students, by spending a year in formal training, save four years' of on-the-job training and experience of the type that could be obtained from singing with small opera companies; this was how the older generation of singers trained in opera (and it is still a route into the profession for many of the present generation). In the next sections of this chapter, we look at the costs of formal training in institutions of higher education and later at the costs of informal training.

The economic analysis of training throws considerable light on how and why the provision for training singers works as it does. By distinguishing general and specific training of all types, it explains why the division between vocational and academic training has no economic function, while showing why the focus of training singers has shifted away from the workplace to the higher education sector.

The Direct Costs of Formal Training

The easiest type of training to put a figure to is training on the private market. The direct costs of private training are fairly easy to calculate, since they consist of the outlay on singing lessons, coaching, and masterclasses and on music, books, records, and suchlike. Taking practice in music colleges as a guide to the number and frequency of lessons needed, that suggests singing lessons of an hour a week for, say, forty weeks of the year for four years. The survey of singing-teachers, showed that the average charge for singing lessons was £14 in 1989. However, a higher fee should be chosen when discussing lessons for professional trainees, since this average fee covered the whole market from children and amateurs to working professional singers; £20 a lesson would be more realistic. Thus the outlay needed for singing lessons at this rate would have been £3,200. Making allowance for an annual rate of inflation of 10 per cent over the four-year period, a trainee starting in 1989 would spend about £3,700 by 1992. Fees for coaching were similar to those for singing lessons but coaching is probably needed less often and only during the later part of the training period. Masterclasses and summer school courses vary in price; such courses typically cost £200 in 1989. Music is also expensive. These items would be likely to add at least another £1,500 to the outlay on singing lessons. Let us boldly assert that it would cost about £5,200 over four years, an average of £1,300 a year for the direct cost of formal training via the private market.

This, however, represents the most basic type of training in singing compared to what is offered in music colleges. No account has been taken of all the other lessons in musicianship, languages, stagecraft, etc. which constitute the course of training for singers in music colleges, though, as we saw in Chapter 1, not necessarily courses in other institutions of higher education in which degrees are offered with singing as a subject in the final examinations. It is precisely such qualitative differences that make it so difficult to compare the cost of training singers in different types of institutions. What made the cost comparison even more difficult in 1990 was that there was such a hotch-potch of funding arrangements for the music colleges, and considerable changes were also being made in the finance of higher education in general. What was the cost of a place in a music college? It would seem to be a simple enough question on the face of it, but in fact it was far from simple to answer. Similarly, in order to answer the question how much more (or less) did it cost to train a singer in a music college rather than in a polytechnic or other institution of higher education it was necessary to grapple with the complexities of the finance of higher education. The first step is to consider what is meant by the term the 'cost of a place' in an institution of higher education.

The Costs of a Place in Institutions of Higher Education

The direct costs of provision consist of staff costs of teachers, the maintenance and rent of buildings, and outlays on library materials and equipment. The total outlay of all institutions within a certain sector—say universities—divided by the number of full-time equivalent students in the sector—all university students—is known by the education authorities as the unit of resource; that is the average outlay per student. The unit of resource in universities has always been higher than the unit of resource in polytechnics and Colleges of Higher Education. Music colleges do not, however, fit into this binary division since only some of the music colleges form part of this scheme of finance. But the unit of resource bears no relation to the economist's notion of cost. For one thing, it averages together students on courses which have very different teaching resource requirements—chemists with laboratories, mathematicians with computers, musicians with practice rooms, etc. The problem is the same even within the fairly narrowly defined area of music teaching; for example, music theory has very different teaching and accommodation needs from organ lessons. In the averaging process, one loses sight of the actual costs of producing the individual product—an hour of music theory or an hour of organ lessons. These distinctions must also be extended to teaching methods; for example, teaching instrumental or vocal technique is done on a one-to-one basis in the form of an hour's individual lesson and requires suitable teaching studios. What is needed, therefore, is information about the costs of particular courses.

However, economists want to go further and distinguish the average cost per student on a course from the marginal cost. The marginal cost is the addition to total cost that comes from taking on one more student; an additional student may increase running costs or capital outlays; one more singing student requires one more hour of singing-teachers' time and the marginal cost of that is easily measured; one more singing student might also mean that another teaching studio has to be built—after which it may be economic to use that to capacity and so indicate that several more students could be accommodated without extra capital outlay. The relation of average costs per student to marginal costs indicates the number of students that it is economically efficient to admit to courses; if marginal costs are below average costs, the number of students could be increased with advantage, and *mutatis mutandis*, if marginal costs are above average costs, this indicates that it would be more efficient to cut student numbers.

There are difficulties in actually working out marginal costs; however, their importance was implicitly recognized in changes that took place at the end of the 1980s in the method by which the UFC and the PCFC, the bodies then responsible for disbursing government subsidies to higher education for, respectively, the universities and polytechnics and Colleges of Higher Education, allocated funds to individual institutions. Whereas until 1989 the unit of resource was used in combination with student numbers as a formula for the allocations, in 1989 institutions were assigned a certain number of resource-funded students and then invited to make bids for the number of additional students they were willing to take at guide prices set by the funding body, these prices being the assumed cost of the place. This system (which is only very briefly touched on here) was supposed, among other things, to get the institutions to work out their own most economic size of student numbers on different courses; institutions which had strong economies of scale or were particularly efficient were further invited to bid for additional students at a price lower than the guide price. In universities, the provisional guide price for Creative Arts courses, which included music, was £3,300 in 1989/90; in polytechnics, a similar concept to that of the guide price was £3,365 for Art, Design, and Performance Arts, which included music. These figures represented an average or target average cost, and it was expected that successful bids would be below this; institutions therefore needed to work out the relation of their marginal costs to average costs.

These changes in the funding mechanism in higher education aimed at increasing efficiency in the production of graduates. A further development in this move to make higher education more responsive to market forces was that fees were raised so that they covered a higher proportion of the cost of a place than had hitherto been the case. With the exception of fees for overseas students, which were always supposed to be cost-covering, fees in the past had been low in relation to the assumed cost of a place. The 1989 changes provided an incentive to institutions of higher education to take in more students where they could 'make a profit', i.e. where the fee covered the marginal cost the students impose. The undergraduate tuition fee in 1990/1 was £1,675 (increased from £607 for 1989/90) and this was the same in both sectors.

To the outlay on tuition fees paid for out of public funds must be added the value of the student maintenance grant, the full amount of which was £2,500 in London in 1989. Taking that as the cost of maintaining a student (whether or not the student actually got a full grant or parental support), and using the guide price of the cost of a place in the Creative Arts or a Performance course and adding in the fee, we see that it cost approximately £6,400 per year to train a singer in a university, polytechnic, or College of Higher Education in 1989/90. With courses lasting three or four years, the total direct cost of training a student in one or another of these institutions of higher education was therefore between £19,200 and £25,600 (in round numbers) in 1989/90.

The Music Colleges

In 1990 the situation of the music colleges was complicated and only some of them fitted into the scheme of funding higher education outlined above. None was funded as part of the university sector, several colleges were an integral part of the public sector of higher education either on their own or as part of a polytechnic or College of Higher Education; some were funded by local authorities and others were essentially private organizations which received grants; one, the London College of Music, was private and the only public money it received came from local authorities via students' fees.

The position for the main music colleges in 1990 was as follows: the Royal Northern College of Music was a monotechnic funded by the PCFC; the Birmingham Conservatoire was part of Birmingham Polytechnic and hence was funded by PCFC; the Royal Scottish Academy of Music and Drama was a Scottish Central Institution and so financed by the Scottish Education Department; the Welsh College of Music and Drama was funded by South Glamorgan Local Educational Authority. The Royal College, the Royal Academy and Trinity College, being old established private institutions, existed mainly on fee income and donations, etc. until 1945, when they began to receive some Government grants; in 1974/5 the DES undertook to fund any deficits that existed between the colleges' income and budgeted expenditures previously agreed between the DES and the colleges: as such deficits persisted, the DES in effect subsidized them regularly. Responsibility for these three colleges was transferred in 1989 to the PCFC and their future was reviewed in 1990 by the Gowrie Committee (see below). The Guildhall School of Music and Drama was set up and continued to be funded by the City of London Corporation and received no central government subsidy, though its students were eligible for local authority awards and so fees were paid out of public funds. Several other colleges, schools, and institutes of music were publicly funded—Dartington College of Arts was a College of Higher Education, Huddersfield School of Music was part of Huddersfield Polytechnic, the Colchester Institute of Higher Education was home to the School of Music; these all fell under the PCFC and/or local authorities.

The Relation between the Cost of a Place and Fees in Music Colleges

Whenever the fee income per student is greater than the marginal or extra cost of a place, an institution has the incentive to take in more students, perhaps even by lowering quality. In the past, this happened in the music colleges. It is important, therefore, to find out the relation between the cost of a place and the fee. But the plethora of funding arrangements and varying combinations of public and private sources of finance led to differences in the relation between student fees and the cost of a place in the different sectors, and particularly in the music colleges. In 1989 undergraduate fees at the Royal College, the Royal Academy, and Trinity College were £1,455, those at the Guildhall and the Royal Scottish Academy were £2,625, and the London College of Music £2,266; in the other music colleges fees were the same as in polytechnics and Colleges of Higher Education. Until the fee was raised, it could not have covered the marginal cost of a student getting one-and-a-half hours a week of individual tuition on first- and second-study instruments or voice in 1989; since the average rate of pay of singing-teachers in higher education was £15 an hour, over the thirty-five weeks of the academic year singing lessons alone cost £525 even without the cost of academic studies. However, it seems possible that the higher 1990/1 fee could have covered marginal costs. Moreover, even in 1989 marginal costs could have been covered in those music colleges charging higher fees.

If the marginal cost of an extra student exceeded the average cost of the place, the incentive was there for colleges to increase student numbers. There is evidence that institutions of higher education do respond to financial incentives; since overseas students have had to pay the full cost of the fee, many universities and polytechnics have actively sought to increase their intake of overseas students. Fees for overseas students in music colleges certainly covered costs, as indeed they were supposed to; here again there was considerable variation; some colleges charged the same fee for all courses—£5,000 in 1990—but others charged different fees for different courses, the highest being £8,100 for the advanced course in singing, including opera, at the Royal Academy of Music (see Table 5 in Appendix 1); there was therefore a strong financial incentive for colleges to take overseas students. The question is whether these figures represented the true cost of a place in the music colleges.

The standard undergraduate fee in 1990 was set at £1,675 by the government with the express purpose of offering institutions of higher education an incentive in the form of extra fee income to raise student numbers. If, as seems at least possible, this fee was above the marginal cost of a student place in singing there would be an expansion of student numbers studying singing. Yet there has long been concern about there being too many places for singers and instrumentalists in music colleges on several grounds: (1) that there is an insufficient supply of suitably gifted entrants, (2) that resources for both teaching and accommodation are inadequate in the music colleges and not up to the standards found elsewhere in higher education, and (3) that the output of musicians from music colleges is too great for the market to absorb. This longstanding concern resurfaced in a report on the Royal College, Royal Academy, and Trinity College commissioned by the PCFC from a Committee of Enquiry chaired by Lord Gowrie (Gowrie 1990). This provided a particularly useful source of information about costs and the finance of these three music colleges.

The Gowrie Report

The report, Review of the London Music Conservatoires (somewhat mistitled, since it dealt with only three of the five) was commissioned by the PCFC when the administration of public subsidy to the Royal College, the Royal Academy, and Trinity College passed to it in 1989. Both the Gilmour Jenkins Report (1965) and the Vaizey Report (1978) had argued for cutting places in music colleges and had advocated a merger between the colleges. A further stimulus to PCFC interest was no doubt provided by the fact that the funding per student in music colleges was very considerably higher than elsewhere in higher education (apart from medicine) and much higher than the £3,365 guide set by the PCFC for Performance courses in the public sector. This report brought to light firm evidence on the direct costs of training singers. It also revealed some extremely fuzzy thinking on the subject of costs both in terms of how to calculate them—presumably the notes on Course Costs in the appendix were written by the PCFC and not by one of the music colleges—and as to how they should be used in decision-making.

The deal that the PCFC offered the Royal College and the Royal Academy (Trinity College at the time opted for moving to Bristol and thereby dropped out of the London question) was that it was prepared to increase funding per student, allowing the colleges to standardize tutorial provision for the principal study instrument at one-and-a-half hours weekly on a one-to-one basis and provide courses lasting four years for a reduced total of students in a newly formed single institution. The Royal College and Royal Academy would be immediately removed from the competitive bidding for PCFC funds (the carrot). Failure to co-operate in forming the new conservatoire would mean that the PCFC could not guarantee the level of funding per student to the separate Royal College and Royal Academy because they would then be exposed to competitive bidding (the stick). As the Royal Academy received 63 per cent of its income in the financial year 1989/90 from the PCFC and the Royal College 66 per cent, that had to be carefully considered (though it was not as much of a threat to them as it would have been to other music colleges, polytechnics, and Colleges of Higher Education funded by PCFC, whose grants represented 85 per cent of their incomes).

The move to federate the Royal College and the Royal Academy would hardly have increased the total outlay by the PCFC, since the proposed 11 per cent reduction of students could allow a 16 per cent increase in the nominal funding per student (a 6 per cent rise in real terms with the rate of inflation in 1990 running around 10 per cent). Thus funding per student could rise from £6,400 in 1990/1 to £7,450 in 1991/2 in nominal terms.

The stated longer run benefits of this federation were to use resources more efficiently, avoid duplication and minimize costs. These phrases have become somewhat platitudinous but what did they mean? When it came down to it, the only costs that could be saved, according to the report, were those for administration and certain central resources such as concert halls and libraries. This seemed to be sensible on the face of it, but nowhere was the anticipated saving of average fixed costs made clear. Nor were the possible extra costs of this sharing (e.g. travelling between sites) shown to be outweighed by the potential savings on fixed costs. Therefore the real financial justification for this move was never made. Furthermore much time was spent on showing how high marginal costs in the form of one-to-one tuition were, though this surely weakened the case for being able to make overall savings. Without knowing the relation of marginal costs, which were predicated as rising, to average costs for different configurations of student numbers there could be no presumption that expanding or contracting student numbers would minimize costs. In the event the recommendations of the report were not implemented in 1990, though they were being reconsidered at the time of writing.

The unit of resource per student place does not represent the whole amount of public subsidy. To the figure for the funding per student must be added the amount of the fee, paid by the local authority for undergraduate students with a local authority award (mandatory for degree courses and discretionary for diploma courses that are not degree equivalent)—but otherwise by the student.[3] This put the figure of funding per student that the 'new' college would receive as income in 1990/1 at £6,400. But figures on subsidy per student do not represent the true cost of a place. Some courses cost more and some less and it was the average over the whole institution that was the basis for the figure of the funding per student. The Gowrie Report also included some figures on the cost of a place by type of course, which provided a very good picture of what formal training of singers cost, at least at the Royal Academy, which apparently had the most sophisticated accounting system of the three colleges. In Annexe 3 to Appendix A of the Report (Gowrie 1990) unit costs were given, measured both using student hours as the basis of apportionment (seemingly the PCFC's recommended method) and using staff hours as the basis (a better method, since it implied that the staff time, which is a cost of production, was being valued—and possibly economized—rather than student time, which is free to the music college). The figures are reproduced here in Table 6 in Appendix 1.

---

[3] If the Royal Academy is anything to go by, the only music college for which there was information on this, three-quarters of performers' course students and all but one of the undergraduates in 1988/9 had local authority awards.

Table 6 shows that, depending on the course and on which measure was used, unit costs at the Royal Academy in 1988/9 ranged from £5,708 to £7,691. It confirms that opera was more costly than other singing courses; with reference to instrumental courses (for which figures are not given here), the Report also showed that singing courses were nearly always more costly than instrumental ones of the same level/type (e.g. Advanced, GRSM, or Diploma). Nor was the Royal Academy the most expensive of the music colleges; overall, the Royal Academy received less funding per student in 1990/1 than the Royal Northern College and the Royal Scottish Academy (Gowrie 1990: para. 4.34). Furthermore, in its submission to the Gowrie Report, the Royal College reported (in Annexe 3 to Appendix C) that the cost per student in Opera Training, a postgraduate course, was £8,345 in 1987/8; this was higher than the subsidy per student at the National Opera Studio at that time.

It is precisely to make this type of cost comparison that a proper calculation of costs is necessary. The Gowrie Committee is to be congratulated on asking for this type of information, and the Royal Academy even more so for being able to produce it in such detail. If it can be assumed that a course at one institution is of the same quality as that at another, it is cost effective to choose the course with the lowest unit costs, providing that unit or average costs are not above marginal costs. This is a correct usage of cost-effectiveness analysis as a guide to the allocation of taxpayers' money for subsidy. But even if it is safe to assume that marginal and average costs are equal, it does not necessarily follow that courses could be merged without raising unit or marginal costs. For example, say there were a situation in which Music College X has unit costs on its postgraduate opera course with an intake of twenty students which are less than those at Music College Y with an intake of ten students. X's costs per student might be lower than Y's because it reaps economies of scale from having more students. But if Y's students were transferred to X this could cause costs per student in X to rise because numbers are now too high to be efficient. The way of finding this out is by comparing marginal with average costs. Therefore simply going for the lowest unit cost is not always a safe strategy for the PCFC or a federated Royal College and Royal Academy to adopt.

The PCFC introduced competitive bidding to get institutions to work out detailed costing of this type. The Gowrie Report offered to remove the Royal College and the Royal Academy from this process if they federated and, taking a narrow economic viewpoint, that seems to have been a generous offer. The subsidy per student in the music colleges was almost double that for a polytechnic creative arts course or a university music course; furthermore, the cost per student was proportionately higher by virtue of the fact that courses mostly lasted four years and not three as in universities (see Chapter 1 for the length of courses at polytechnics). On the other hand it cannot be said that the course quality in the training of instrumental and vocal performers was everywhere the same. Under the 1990 funding arrangements, student numbers would probably nevertheless be induced to expand in music and performance courses outside the music colleges.

The Gowrie Report was unfortunately unable to get to the nub of the problem of the unco-ordinated production of music students here and there over the whole spectrum of institutions in higher education. Because of this its recommendations were made on a partial basis. Gowrie and his committee probably were not able to widen the scope of their enquiry and certainly showed themselves aware of the need for wider co-ordination. But the PCFC should have known better and not have put them in that position. The PCFC was responsible for several music colleges and for other polytechnics and Colleges of Higher Education which specialized in music. It was right to look at provision in London—though even there the review was not comprehensive—but wrong to look at it in isolation from the rest of the country. Eventually there must be a debate that encompasses total provision for higher education in music (and perhaps gives more attention to singers as a sub-group with particular needs).

Summary of the Analysis of the Direct Costs of Formal Training

Earlier, a rough estimate was made of the cost of private formal training in singing of £5,200 for four years. This figure did not include the cost of student maintenance. The cost of training an undergraduate singing student in 1989/90 in a university, polytechnic, or College of Higher Education, including the cost of student maintenance (which may or may not have been paid out of public funds) was between £19,200 for a three-year course and £25,600 for a four-year course.

What was the equivalent cost for an undergraduate singing student in a music college? This is more difficult to assess, because of the considerable variation in the cost of places and fees in individual music colleges. The Gowrie Report provided evidence on this; it offered the merged Royal College and Royal Academy £6,400 per student (including the standard fee) for 1990/1, which is one figure that could be used in this comparison; it also stated that the funding per student in the music colleges that it was supporting at the time was between £5,162 and £6,225. For the sake of argument, let us use a round figure of £6,000 (including the fee) for the calculation. Again, adding the cost of student maintenance of £2,500, this made the direct cost of training an undergraduate singing student for four years £34,000 using the 1990/1 figures.

The Indirect Costs of Formal Training

The indirect costs of formal training in full-time courses in higher education consist of the opportunity cost of forgone earnings during the period of study. These costs are calculated with reference to income from alternative employment that a trainee could have undertaken had he or she not made the decision to become a student. They therefore depend upon the length of the formal training period, the age and sex of the trainee, and the educational qualifications that the trainee had at the outset of training, since earning power generally depends upon age, sex, and educational achievement.

A singing student with A-levels starting a four-year course at the age of 18 forgoes four years' income of the average for 18–22-year-olds in another occupation for which A-level is the normal entrance qualification. That, of course, is the case for any student in any discipline in higher education. The logic behind this application of the concept of opportunity costs is on the one hand, that output is lost to society and, on the other, that income is not enjoyed by the student. The former is the social cost of higher education and the latter the private cost. Which notion should be used depends upon the nature of the enquiry; if we want to know whether society benefits from training more singers rather than economists or bank clerks, the social cost would be the appropriate cost, but if the question is whether an individual would be better off investing in training as a singer instead of becoming an economist or a bank clerk, private cost is the relevant measure. These measures are discussed in Chapter 5; the present task is to establish what the indirect costs are of training in general.

The New Earnings Survey 1989 (Department of Employment, 1989) calculated non-manual gross weekly earnings by sex and age: the figures are given in Table 7 in Appendix 1. These figures show that a singing student beginning a four-year singing course at the age of 18 in 1989 experienced an indirect cost of £32,604 if male and £28,169 if female over the four years. (These figures do not take account of wage inflation over the four-year period.) If entry was at the age of 20, the indirect cost for a male rose to £40,196 and £32,381 for a female. The longer the study period lasts and the later it begins, the higher the indirect costs of formal training become.

For advanced singing students, such as students at the National Opera Studio, who are typically in the 25–9 age range, one year at the Studio meant an indirect cost of £14,388 for a man and £10,722 for a woman (any earnings in the period to be deducted from this figure). In 1989 the fee for the course was £4,000; comparing these figures dramatically conveys the significance of the indirect costs of training and substantiates the claim that a major part of the total cost of training singers must be borne by the singers themselves despite the considerable contribution made by public subsidy.

The Cost of Informal Training for Singers

As explained at some length earlier in this chapter, informal or on-the-job training is a vital part of the process of training singers; at some point learning can only be achieved by doing; in the final analysis, performing can only be learned by performing in public. Because much of the skills of a performer are of a general nature, the cost of acquiring informal training is passed on to the trainee and takes the form of low earnings. In a freelance labour-market, such as exists for singers, musicians, and other performing artists, low earnings result from both low performance fees and a low number of performance opportunities. Many singers between the ages of 25 and 30 who are concentrating on solo work would be glad to average thirty performances a year during those five years, earning in 1989 probably around £200 a performance. Here again the notion of opportunity cost is invoked; had the singer decided on graduation to become a teacher—an option open to many who have taken a graduate course—he or she would have been earning about £10,000 a year on average during those years. Thus the cost of informal training using these fairly realistic figures was 5 years × £10,000 = £50,000 that could be earned in teaching minus the singer's earnings, namely 30 performances at £200 performance fee × 5 years = £30,000. Thus the cost of informal training while the singer gained experience and trained on the job was £20,000 over five years, assuming that the singer did no other paid work besides singing during those years.

Search Costs

Besides the opportunity cost of forgone earnings during both the formal and informal training period, singers also experience costs of looking for work. These are search costs. All workers have to look for work at various points in their careers but the costs for singers are particularly high as the labour-market they work in is predominantly freelance. In other professions, the job search involves looking at job advertisements, writing letters to prospective employers stating qualifications and work experience, and attending interviews. In the singing profession, few jobs are advertised, paper qualifications are largely ignored by employers, and selection for each job is done on the basis of an audition. Young singers therefore experience considerable costs in attending auditions. Entering competitions is one way of saving the cost of searching for work, since it allows the young singer to be heard by a number of potential employers, but it costs money to enter competitions and this must be regarded as a search cost. Music colleges also assist this process by offering singers public performances thereby saving search costs. Once a singer has an agent, the agent will (or should) undertake the search on behalf of the singer; saving on search costs is the chief economic role of agents.

It is very difficult to quantify search costs; however, interviews with singers suggested that they were between £1,000 and £2,000 in 1990.

The Cost of Training Singers and the Supply of Singers

In Chapter 1, it was argued that the output of higher education courses in which singing forms part of the final examination is not a measure of the supply of singers as there is no close correspondence between what students study and how many seek to enter the singing profession.

This is equally true of all other courses in higher education. First Destinations data on recent graduates collected by every university, polytechnic, and College of Higher Education in Britain show this to be the case. For example, only 6 per cent of economics graduates got a job with the job title 'economist' in 1989. Does this mean that resources have been wasted in training graduates to do one thing if they go off to do another? Not really, because labour-markets for most graduates recognize graduate training as being general rather than specifically vocational. Having a formal training in singing which leads to a recognized qualification—a degree or diploma—provides a singer with the means to enter the general graduate labour-market. This view was also expressed by Vaizey (1978: 9). In this sense, it cannot be said that public money spent on training singers is wasted if graduates and diplomates in singing do not become singers, any more than it could be said of economists or philosophers; all benefit from higher education.

But because it costs more to train singers than other graduates the question does arise as to whether the extra public money spent on their training is well spent if they do not become singers. If after four years' relatively more expensive training a singing student takes employment in a field that he or she could have entered with three years of less expensive training, one might be tempted to say that the extra public money involved had been wasted.

That it is more expensive to train singers in music colleges than other graduates can be seen from UFC and PCFC figures of guide prices for the competitive bidding process, for the 1990/1 academic year. A selection is given in Table 8 in Appendix 1; they ranged from £2,800 a year for Humanities to £4,600 a year for a university Engineering course.

The Gowrie Report stated that the overall figure of funding per student in polytechnics was £3,866 in 1990/1, as compared to between £5,162 and £6,225 in those music colleges funded by the PCFC. Thus in 1990 four years at a PCFC funded music college cost between £20,648 and £24,900 as opposed to £13,251 for three years in a university music department, £18,468 for a four-year Creative Arts course in a polytechnic, or £10,833 to train a Business Studies or similar graduate at a polytechnic (these calculations include a standardized fee but exclude the cost of a student maintenance award.[4] Quite why unit costs in music colleges were so much higher than those in other institutions of higher education with courses in the performing arts and music is not clear, particularly as the latter often offered equivalent individual tuition on a one-to-one basis. A possible answer is that comparable courses in other institutions have not been properly costed. But whatever the reason, there was a considerable discrepancy.

But those costs are the direct costs of formal training. The true costs of training singers are even higher than those of training other professionals because, in addition, there are high indirect costs, on-the-job training costs and search costs. On top of that, singers have to bear the cost of singing lessons and coaching throughout their career. Undergraduate courses (in any type of institution) offer sufficient professional training for only a small minority of students of singing; most require further study as postgraduates before they are ready to enter the singing profession. It is at this point that public funding ceases to support students in Britain, since there is no postgraduate award scheme for singers. Unless they can get a bursary from a music college or charitable foundation, singers have to pay for advanced training themselves. The exception to this is the National Opera Studio, as explained above.

Postgraduate singers are typically older than their counterparts in other disciplines and they may have to spend several years preparing for entry into the singing profession. The older the student is and the longer training takes, the higher the indirect costs of training in the form of forgone earnings. Even when 'young' singers, often in their late twenties, start to work professionally as soloists they do not get much work and earn relatively little from it. However, those singers who go into choirs and sessions work with little postgraduate training avoid these extra indirect costs and search costs; choral scholars in particular seem to take this option.

Why is there no sysyem of grants for postgraduate singers? They are surely a better investment than undergraduate singers; the music colleges know much more about their ability than they do about undergraduate entrants and by that stage the students are highly motivated. It seems perverse that the government subsidizes many

[4] The Gowrie calculation implied a fee of £1,117 being a figure somewhere between the fee levels of the 1980s and the fee of £1,675 that was to be introduced for 1990/1.

undergraduates who will not make the grade as singers but not the more able who have made an informed career decision. There is an economic explanation for this apparent perversity. The government subsidizes higher education in order to promote a supply of graduates with general training on the one hand, and on the other, through postgraduate studentships, it ensures a supply of highly trained specialists. The pattern of undergraduate provision is demand-led and it would be considered inequitable and dirigiste to withdraw subsidy from specific courses and disciplines (though some of this has taken place, it was largely a result of decisions by individual institutions). At postgraduate level, public support can more easily be directed to specific disciplines which the government wishes to support, because, rightly or wrongly, it believes they are essential to economic growth. Singers are neither regarded as a priority, nor is there a shortage of them.

Despite the high costs of training that singers experience, there is a plentiful supply of new entrants to the singing profession. Evidence of this is to be found in the pressure of applications for places on courses for singers of all types and at all levels, and in the large numbers who audition each year for opera companies and conductors, who write to agents, and send their brochures out to all and sundry. There seems always to be an excess supply of singers for all types of work even at low rates of pay; indeed, singers will work for nothing, and not just when they are young. Supply does not seem to be affected by the cost of training, probably because students regard themselves as making a good investment; they overvalue their chances of success and are not well informed about the difficulties of earning a living as a singer. These issues are discussed in Chapters 4 and 5.

Student grants which largely cover the cost of undergraduate study encourage people to train as singers, just as they encourage students of all disciplines to enter higher education. Whether this leads to an oversupply of singers is a complex issue that can only be answered in relation to the demand for singers and in relation to earnings.

Part II
The Demand for Singers

3
Employment in the Market for Singers

Like supply, demand is a relationship between price and quantity—here between the number of singers or performances per singer that would be employed at various fees or wages. We expect the demand relationship to be inverse—the lower the fee or wage rate the more labour will be demanded. As with supply, this is a difficult concept to measure and we end up measuring demand by employment. Employment can be thought of as a stock, the number of jobs in existence at any one time, or a flow, the number of people hired per period of time. This chapter deals with the demand side of the market for singers, describing opportunities for employment in Britain offered by the various organizations which promote opera, oratorio, concerts, etc. as well as in recording and broadcasting.

The question of employment may be viewed from two points of view; how much employment is on offer to singers in aggregate and, from the singer's point of view, how much work constitutes full employment. This distinction arises because of the freelance nature of the labour-market for singers. In a labour-market in which regularly contracted work is the norm, the correlation between the number of jobs, i.e. the amount of employment and the number of workers employed, is strong. For example, in the teaching profession the number of posts for teachers and the number of teachers employed are known; some teachers work part time and their part-time hours can be calculated in terms of full-time equivalent posts. At any point in time it is therefore possible to state the amount of employment available to teachers and, taking account of unfilled posts, teachers on leave, and so on, we could calculate how many teachers are employed. If more posts exist than teachers employed in them, we could say that there is a shortage of teachers. In such a labour-market there are norms as to what constitutes full-time work; the employment contract states so many hours a week must be spent in the workplace and lays down holiday time and other such conditions of work.

However, where workers are self-employed this type of easy calculation is not possible and norms about hours of work and holiday periods can really only be established with regard to the individual; one singer may regard him- or herself as working full time doing thirty performances a year; for another it could be 130. Because of the uncertainty that all singers, except the stars, experience with respect to employment opportunities, the situation can arise where a singer takes on anything that is offered and finds herself with too much work for one part of the year and nothing for the rest. This is, of course, true of other self-employed people but they do not all have to meet a performance date in front of the public. A further aspect of this problem is that solo singers mostly have to prepare repertoire in their own time; it can take singers months to study a major operatic role or prepare for a solo recital. Though they may in fact be working full time in so doing, there is no objective way of measuring the time taken by the usual concept of hours worked. Therefore, questions about how many singers are employed or unemployed are complex.

In fact, the market for singers can be divided into two sections; that for singers who have regular contracted work and are employed by an organization and the other in which singers are self-employed. This can be clarified by describing employment for singers with reference to the different sectors of the market. The information is based on questionnaires and interviews with a large number of organizations, conducted in 1989–90.

Employment Opportunities

The market for singers in Britain has come to be dominated by opera. Employment opportunities in opera may be classified as permanent, seasonal, and occasional or casual. This partly reflects the nature of the opera companies themselves and partly the differing types of employment they offer. In 1990 there were five permanently established opera companies which performed throughout the year: the Royal Opera, English National Opera, Opera North, Welsh National Opera, and Scottish Opera. All these companies had a permanent chorus with annually renewable contracts; in 1989 they employed 275 singers in total in their full-time choruses. In addition, they employed 118 extra chorus singers on a casual basis for the run of performances of a particular opera (over half of these being at the Royal Opera House). Besides regular employment for the chorus, these opera companies also employed forty-seven soloists (principals) as company artists on a regular contract in 1989, the majority (thirty-nine) being with the Royal Opera and the English National Opera and having renewable contracts. Those employed in this capacity by Opera North, Scottish National Opera, and Welsh National Opera were typically young singers at the beginning of their careers and their contracts were usually for one or two years.

Together, the permanent opera companies employed 373 guest artists to sing solo roles in a total of 641 performances of sixty-five operas in 1989. It is in respect of this employment that difficulties begin in assessing the amount of employment available to individual singers and it would be meaningless to average out these figures, since the situation would be different for each individual. These figures may include someone singing only one performance of a role in one opera and someone else who sang many performances in several operas. (What one would ideally like to know from the point of view of the amount of work available is the number of role-performances per year and, from the point of view of the amount of work undertaken by an individual, the number of performances each person did and then what proportion was done by company or guest artists; because it was deemed unreasonable to ask for this information, the aggregate figures must suffice.)

In addition to work in main-house performances, the permanent companies also employed singers, usually on an occasional basis, to do educational work and Scottish Opera and Welsh National Opera had small-scale touring companies for a short season for which extra singers were hired—about thirty-five all told. Also, singers were hired to cover (understudy) roles; in 1989 240 covers were hired (again, some may have been the same individual). This type of work is mostly of an occasional or casual nature but it is extremely important to young singers as a source of on-the-job training experience.

The seasonal opera companies, Glyndebourne Festival Opera and Glyndebourne Touring Opera, Opera 80, Opera Northern Ireland, Opera Factory–London Sinfonietta, Buxton Festival Opera, employed ninety-eight chorus singers, for between five and eighteen weeks in 1989, thirty-seven company and 114 guest artists for a total of 201 performances of twenty operas given over seasons of between one and fifteen weeks in the year. Again, how much work this could produce for any particular individual is very difficult to say; in the case of the Glyndebourne companies and Opera 80, there could be significant blocks of work for singers over the three or four months during which their seasons last with substantial numbers of performances over the season. In other companies the season was only two or three weeks. Certainly for young singers, Glyndebourne chorus was an important training ground and offered the chance for gifted young singers to cover roles during the Festival and do a solo role during the tour. Similarly, Opera 80 offered the opportunity for young singers to perform principal roles in smaller venues. The same used also to be true of Kent Opera and its demise was much to be regretted as a training-ground for young singers.

Besides these well-established companies, there were many smaller opera companies which offered employment opportunities of an occasional kind to singers at various levels of experience; the majority employed young singers but some had well-established singers singing main roles at times.

An attempt was made to discover the extent of the demand for professional singers on the part of these smaller opera groups. Under this heading were Travelling Opera and Pavilion Opera, companies regularly touring with professional singers, chamber opera groups, big semi-professional companies (Kentish Opera, Dorset Opera, and the like), which hired professional principals to work with their amateur choruses, and *ad hoc* groups of all types. The problem with these smaller opera groups was that they tended to be ephemeral and it was very difficult to trace them; I attempted to do so by a postal questionnaire but many of my enquiries were 'returned to sender, not known at this address'. In the end, I got responses from seventeen groups which hired professional singers and information about the number of professionals hired and how much (if anything) they were paid from ten of these.[1] These results cannot therefore be regarded as a sample survey since I have no figure for the total population of these groups from which these responses were drawn; the following information is merely offered as it stands and I am unable to generalize reliably about this type of activity.

---

[1] The information related to their last complete season, which was usually 1988 or 1988/9, depending on how the group was organized.

The smaller opera groups were of interest on two counts: (1) they offered performing experience to young singers in the form of on-the-job training and (2) they gave some insight into the semi-professional world inhabited by quite a large number of trained singers, who, for one reason or another, have not made a professional career. As far as one can tell from biographies of the principal singers engaged by these groups, they had all had a substantial training in singing either at music colleges or with private teachers.

Several of these opera groups offered singers fairly substantial employment opportunities. Travelling Opera had done between forty and eighty performances a year since it was formed in 1987. It toured to small venues (though it has also had short seasons at Sadlers Wells) with a small orchestra and twenty to twenty-five singers. Pavilion Opera toured stately homes with about twenty singers and performed with piano accompaniment; in 1988 it had a repertoire of six operas and gave sixty performances in Britain and twenty-three abroad in a forty-four-week season. London Opera Players and London Chamber Opera gave thirty performances of a repertoire covering four full-length operas and nine one-act chamber operas, employing thirty-nine professional singers in 1988/9. On the other hand, other groups gave only a few performances of one opera in the 1988/9 season. All told, these seventeen smaller opera companies offered employment to a total of 126 singers and gave nearly two hundred performances.

It can be seen from this that the smaller opera groups offered singers performing experience and in some cases they offered the chance of earning reasonable amounts of money while doing so (see Chapter 4 on Earnings). Thus these groups made a significant contribution to opportunities for professional singers to gain experience of playing major roles in standard repertoire operas, sometimes with orchestra, to paying audiences. Many of the older generation of singers in Britain gained their early performing experience in this way. In the 1980s these small opera companies were in decline. Their decline was attributed partly to economic reasons and partly to the higher standards of the quality of performances increasingly demanded by a more informed opera-going public, particularly in provincial areas. However, there seemed to be a resurgence of this type of activity in the early 1990s.

For reasons given earlier, it is impossible to chart the rise and fall in the number of smaller opera groups; they mushroom into existence and fade away for a variety of causes. Many depend upon the dedication of an individual who moves on; the costs of running these groups, which have to depend largely on box office takings, can easily become prohibitive. Nevertheless, they continue to appear as well as to disappear and provide a useful platform for both aspiring young professional and experienced semi-professional singers.[2]

The groups providing information for this survey were generally speaking the larger and better organized ones with a record of reasonably high performance standards and their performances were often reviewed as 'Fringe Events' by *Opera* magazine. It should be noted, therefore, that the picture given here is of the better opportunities offered to singers by this type of activity. For that reason it would be inappropriate to attempt to generalize these findings to all such groups, even if they could be traced. Besides these professional and semi-professional smaller opera groups, amateur opera groups also offered unpaid performance opportunities to young singers, but it proved impossible to gain any systematic information about them.

Commercial Promoters

By 1990 there had been several spectacular large-scale productions of opera by commercial promoters, the productions at Olympia being the biggest undertaking. For the 1989 production of *Carmen*, which subsequently went on a world tour, a large chorus of ninety singers was hired for several months as well as top-ranking opera soloists. There were other commercial promotions of excerpts of operas, choruses and arias, which offered employment to professional chorus singers and soloists. As opera grows in popularity, one might expect this type of employment opportunity to increase.

Musicals also employed trained singers, some of whom had received a classical training; *Phantom of the Opera* was one such and also *King* before it folded. Generally speaking, musicals require singers who can make a 'pop sound' using microphones rather than the usual projected sound that college-trained singers learn to make, and singers who can dance as well—hence musicals employ rather differently trained singers than the normal classical training produces.

[2] As this book was being prepared for publication, a report on small opera companies by Graham Devlin was being written for the Calouste Gulbenkian Foundation, which should provide more recent information on this source of employment.

Nevertheless, some college-trained singers do get work in musicals and colleges are becoming more aware of the need to train singers in this field.

Concert and Sessions Work

There was a wide range of work in the concert and sessions sector of the labour-market for singers, though, because it was almost all contracted on a freelance basis, it was impossible to quantify either the amount of employment or the extent to which individual singers worked in it. In this sector employment generally stems from the different promoters—orchestras, fixers who gather choirs together for specific jobs, and choirs which, though they employ freelance singers by the session, nevertheless draw on particular groups of singers on a fairly regular basis.

The one employer which contracted singers throughout the year was the BBC Singers, employing twenty-four singers in 1989; they had an initial three-year contract which was thereafter renewable annually. Singers worked five days a week, usually for thirty hours a week (ten sessions, but they could be required for eleven). The preference was for singers with good solo voices as well as the ability to sing contemporary music at sight, and the singers were encouraged to take on solo work on their own account; this was made possible by having a pool of *ad hoc* singers who could deputize for regular members; this pool consisted of about 250 singers who had passed BBC auditions. The pool could also be drawn on for performances of works that required more than twenty-four singers—the BBC Singers could go up to seventy or so in strength for such occasions and singers from the pool were then employed on an *ad hoc* freelance basis.

Other regular ensembles of singers existed, which, however, did not have annual contracts but hired singers on a session basis (a session being usually of three hours' duration). The Monteverdi Choir was perhaps the best established of this type of choir in the classical field; they had twenty-eight regulars and around fifty other singers on the books in 1989. 'Full-time' work with the Choir in 1989 was for an average of two sessions a day for four months of the year. Like the BBC Singers, soloists were often drawn from the choir and many were likely to go on to solo careers. Other choirs were linked to orchestras which promoted concerts and did recordings, some specializing in one type of music or another; the popularity of Early music and historically aware performance of Baroque, Classical, and latterly early Romantic music has given rise to (and, indeed, has been led by) the development of specialist choirs, such as The Sixteen Choir, the Schütz Choir of London, the Tallis Scholars, Taverner Choir, and Gothic Voices to name some. Other choirs and ensembles specialized in contemporary music, examples being London Sinfonietta Voices, Electric Phoenix, Jane's Minstrels, and Sing Circle. Another configuration was of choruses specializing in opera recordings, such as the John McCarthy Singers and his Ambrosian Singers and Ambrosian Opera Chorus; others specialized in light entertainment, the Swingle Singers being perhaps the longest standing of this type of group.

Each of these choirs, choruses, and ensembles had a fixer or choral manager, the person who got a group together for a job. Equity had a list of thirty-nine Approved Choral Managers in 1989, people who would ensure that Equity rates were paid to singers and that Equity contracts were adhered to. Equity also had a register of concert and sessions singers; in 1989 Equity reported having 1,119 such singers on their concert and sessions list.[3] Though many had a regular affiliation to one or another fixer, most of these singers were available for work with other fixers and went where the work was. The type of work was legion—recording work of all kinds (from opera to pop music), TV, radio ('Friday Night Is Music Night' regularly employed concert singers), films, commercials, jingles, voice-overs, weddings, banquets, hotel work, cruises, backing for pop singers, cabaret, barber's shop quartets—anything that required people who know how to use their voice professionally. Concert and sessions singers go from one job to another and from one type of work to another. Many audiences would be amazed to find out that the soloist in the Bach chorale they just heard had made a TV commercial that morning! Fixers often had up to two hundred singers on their books and could produce a number of expert singers for almost any occasion. A large professional choir in 1990 would be forty singers though sixteen to twenty was more usual. Larger choirs were nearly always amateur, simply because it was too expensive to hire professionals. Singers were often regarded as the 'icing on the

---

[3] This information was supplied by Equity. Other people interviewed, particularly several of the Approved Choral Managers themselves, have said they believed there are anything from 2,000–5,000 such singers. It was difficult to check these figures.

cake', being less 'necessary' than instrumental players and hence the budget for singers was what was left over. How much work a particular singer got in all this was an entirely individual matter; some worked all the time, even specializing in a very narrow field (there were reputed to be ten to fifteen singers who had completely cornered the jingles market, earning a good living from making the tape and from repeat fees); others had some sort of bread-and-butter job—teaching, a church job, etc.—that provided a regular basic income.

Orchestras

Orchestras promote their own concerts of works which require singers either as soloists or choristers. The big orchestras often have an associated amateur choir which they use for the vocal–orchestral repertoire—the London Symphony Orchestra has the London Symphony Chorus, the Philharmonia Orchestra the Philharmonia Choir, and so on; while it is true in one sense that amateur choirs displace work for professional choral singers, it is likely to be the case that without amateurs the orchestras simply could not afford to play this repertoire. I conducted a postal survey of orchestras to find the extent to which orchestras employed professional singers in their own promotions. The repertoire for which singers were used covered opera (concert performances of whole operas, excerpts or arias) masses, cantatas, oratorios, concert arias, song cycles, and individual songs or arias. In a third of the cases in which a choir was used a professional choir (typically of fifteen to twenty singers) was hired, most frequently for Baroque music. The largest professional choir of twenty-four singers that came up in the survey cost over £4,000 for one concert in the 1988/9 season, which gives an idea of the cost of hiring professionals. The average number of soloists hired in the 1988/9 season by orchestras promoting concerts involving singers was fourteen; half the sixty responding orchestras promoted concerts which required singers and, as these were usually the bigger orchestras, this probably means that an estimate based on these figures would very likely overstate the overall demand for soloists in this field. My estimate is that 1,100 solo singers were hired in 1988/9 in a total of just over 1,100 performances of vocal music in own promotions by orchestras. These figures do not, of course, include orchestras' performances with vocal soloists or choirs which were promoted by other organizations such as festivals, so the total employment of singers in this type of repertoire would on that count be greater than they suggest.

Amateur Choral Societies

The amateur performing tradition in Britain is very important for the employment of professional singers. The NFMS is the body which exists to advise and aid the work of amateur music clubs and performing societies, both choral and orchestral. Until 1985 the NFMS, funded by the Arts Council, was responsible for disbursing financial support to its members to enable them to pay for professional conductors, chorus-masters, orchestral players, and singers; this function was then transferred to the Regional Arts Associations in England and Wales, Scotland remaining with the NFMS; not all amateur performing societies were members of NFMS but the majority were. There were around 800 choral societies in 1990 which were NFMS members. They typically put on two performances a year, though the bigger ones did three. Many societies hired professional singers as soloists (and a few hired choristers to stiffen the choir); the biggest of the societies, such as the Huddersfield Choral Society and the Royal Choral Society paid top rates to soloists, while many smaller societies paid so-called student rates or had amateur soloists. With four soloists usually being called for in the oratorio repertoire this implied a demand for 6,400 singer-performances in 1990 (not all of them being with professional singers) with choral societies, and many singers specializing in this repertoire were able to make a good living in this sector of the market. The societies tended to perform at certain peak periods of the year, i.e. Christmas and Easter, and so demand was highly seasonal; they also tended to concentrate on a few works, Handel's *Messiah*, Fauré's *Requiem*, and Mendelssohn's *Elijah* being favourites. (An analysis of fees paid by choral societies to singers from 1985–9 is discussed in detail in Chapter 4.)

Recitals

Opportunities for solo recitals by singers were reported to be declining in the 1980s, by all accounts. Music clubs in the NFMS which promoted concerts no longer favoured singers, though at one time of day this was an important source of employment for song recitalists; very few clubs had a solo recital by a singer in 1990. The BBC was the biggest employer of singers for recital work. A search through the *Radio Times* for 1989 revealed that on Radio 3 there were sixty recitals of which well over 80 per cent were BBC product (rather than those of foreign radio stations), two-thirds were live broadcasts, many being the BBC's own promotion. The BBC regularly auditioned singers (thirty-five in 1989) and maintained a pool of thirty to forty solo singers per voice type from which it chose singers for recitals; these were all experienced singers, not beginners, as only experienced singers could be considered for a broadcast recital. In addition, BBC staff went to performances and competitions to listen and report on singers.

Music festivals offered significant opportunities for recitalists but again, these were usually well-known, often internationally known, singers. For beginners and, indeed, for many experienced singers, the only option for recitals was to promote them themselves. There were various venues where this could be done; several churches in the City and Central London (and in other cities too) had a series of lunchtime recitals which included singers, university and other similar music societies, libraries, fringe festivals were all possible venues as well as the Purcell Room and the Wigmore Hall. Promoting your own recital could be a very expensive venture; a Wigmore Hall recital, including hall hire, concert management, and suchlike, cost about £2,000 in 1990.[4]

Another way that singers were able to do recitals was by forming duos or small ensembles with other singers and instrumentalists (and occasionally with a narrator). The best known of this type of group in the classical recital repertoire was The Song Makers' Almanac or the King's Singers in the lighter repertoire.

Church Choirs

Westminster Abbey, St Paul's Cathedral, Westminster Cathedral, and St Margaret's Westminster all had a regular salaried choir of men with boy trebles. St Paul's, for example, had eighteen men in 1990 whose duties consisted of Sung Evensong every day, Mattins

---

[4] This was the total cost; any revenue from ticket sales would reduce the outlay. There are 540 seats in the Wigmore Hall; if all were sold in 1990 at an average ticket price of £5, the soloist could cover costs out of revenue, though the singer would still have to pay the accompanist and the concert agency out of that. Audiences would probably be astonished to learn that even very well-known soloists shared the promotion of their own concerts with the Wigmore Hall. Without more funding, halls such as the Wigmore could not afford to take financial risks, part of which, therefore, were passed on to the performer.

on Saturdays and Sundays with Sung Eucharist on Sunday as well; the men could have up to half the 'statutory' services off by providing a deputy who was paid directly out of their salaries so that this was an attractive regular job for men wishing to work elsewhere in the market for singers. Weddings, memorials, Guild services, etc. and recordings with the choir were a source of extra income.

Other places of worship—churches of various denominations and synagogues—also regularly hired singers by the session to sing in services and at weddings, etc. For some singers this provided reliable bread-and-butter earnings and they retained their association with a choir over a long period; for others it was an opportunity to earn extra money and was effectively another type of sessions work. The hiring and payment of choir work was usually in the hands of the organist.

Teaching Singing and Related Work

Teaching singing was another source of employment for singers in music colleges, other institutions of higher and further education, music centres, a few schools, and also privately. Many singing-teachers teach part time, combining teaching with a performing career. Even in the music colleges and institutions of higher education which teach singing there were relatively few full-time posts; private teachers worked either full time or part time. This was discussed in detail in Chapter 1.

Other work connected with the training of singers was giving masterclasses, talks, or lectures; acting as a vocal adviser on audition panels and the like; and adjudicating singing competitions. All this type of work was negotiated on an *ad hoc* basis; it could keep some singers busy even well beyond the normal age of retirement.

Employment Contracts and Conditions of Work

An initial discussion of the role of employment contracts and conditions of work was given in the Introduction and the fact that contractual arrangements may affect the demand for singers was mentioned. Employment contracts and conditions of work vary according to the type of employer and the type of work, and are extremely complex; it is not my intention here to go into detail but to say enough to indicate what the economic effects on the market for singers are (any reader who is a glutton for punishment may get copies of contracts and rates of pay from the ISM and Equity). The descriptions relate to the situation in 1990 (hence the use of past tense) but these are long standing arrangements and will probably remain in place for some time to come.

Opera

Contracts and conditions of work in opera were negotiated by Equity in agreement with the Society of West End Theatres, Theatre Managers' Association, and the Theatres National Committee and accepted by all the opera companies. There were two types of contract, the Esher Standard Contract for Opera Singers and the Standard Contract for Guest Artists engaged by Opera Companies; they were similar with respect to conditions of work but applied different payment arrangements: the Esher contract dealt with weekly payments and was therefore relevant to choristers and company singers (Equity uses the term chorister for singers in choruses); the guest artist contract dealt with fee payments for freelance singers in opera. However, each of the five permanent opera companies had an individually negotiated chorus contract in which pay agreements differed, though again conditions of work were very similar. Under the Esher contract, singers were paid by the week for rehearsals and performances alike; under the guest artist contract payment could be either by the week or pro rata for the number of performances or a combination of the two; if the performance fee was above £480 (in 1990) there was no duty for the employer to pay for rehearsal and learning time, but if it was less, then rehearsal time must be paid for as well, along with travel expenses and Touring and Subsistence Allowances if the company was touring or if the singer had to live away from home. Touring and Subsistence Allowances, negotiated annually along with minimum weekly pay, also were part of the Esher contract and travel time as well as expenses had to be paid if a singer had to live away from home. (The market, not surprisingly, recognized this distinction, by producing a gap in fees paid per performance from around £200 to £480 in 1990 when the minimum weekly wage rate was £165.)

Other conditions of work in opera in the Esher contract related to the number of hours a week to be worked; the Monday to Saturday working week consisted of ten working sessions of three hours' duration in weeks in which performances were taking place and eleven in pre-production weeks, there being no more than two sessions a day with a break of one hour between them. There had to be a rest period of eleven and a half hours calculated from fifteen minutes after the singer left the stage. A singer doing a principal part other than a minor role could not be required to rehearse during the afternoon of a performance. A nominated dress rehearsal session might not be longer than four hours on a performance day or six hours with a one hour break if there was no performance that day. Overtime rates were paid if dressing time was required in addition to the four- or six-hour periods, and whenever there were working sessions extra to the ten or eleven weekly sessions or there were Sunday rehearsals, as well as for late performances, where late meant in excess of three and three-quarter hours after signing-in time (no less than half an hour before the singer's stage call). The Esher contract also stipulated a maximum of five weeks' paid holiday for singers on an annual contract and entitlement to a half-day's holiday with pay for every week worked for singers working part of the year. The only statutory holidays on which singers were not required to rehearse or perform were Christmas Day and Good Friday but extra payments or time off in lieu could be given. There were time allowances for choristers putting on and taking off full body make-up and extra payments for a variety of items, singing and learning in a foreign language (worth £11.03 a week in 1989 at the Royal Opera), certain sorts of dancing, moving or carrying things on stage, screaming, speaking a line, and so on. While some of these conditions may seem rather overdone, it has to be borne in mind that a singer must protect the voice from strain and remain in good bodily health to be able to work satisfactorily. In some choruses, there was a vocal maintenance payment to enable singers to have lessons. The details given here are of the Esher contract in 1990 but, as stated earlier, there were local variations in every individual opera company, for example, the chorus at the English National Opera was required to do only nine sessions a week and individual choristers do no more than 110 performances in the year; some choruses had six weeks' holiday not five, and so on.

These conditions obviously had implications for the scheduling of performances and rehearsals and even the choice of operas. Long operas, for example those by Wagner which involve large choral forces, required not only considerable extra payments but also rehearsal and performance schedules that provided sufficient rest periods. In turn, the choice of operas determined the extent of opportunities to earn overtime payments; some operas only require a male chorus (apparently these were more frequently performed than ones requiring only a female chorus) with the result that men choristers earned more in overtime than women; some operas require a small chorus so only part of the chorus would be needed; the choristers chosen for the small chorus thus had the chance to earn more if overtime was involved. The resulting imbalance of earnings between choristers could be a source of friction within choruses and between choruses and opera managements.

Another factor that affected choristers' employment and earnings in other ways was the rehearsal and performance schedule. The management was required to give notice of the next week's schedule only on the Friday of the week before, so that singers knew when they would have free time. This determined whether or not they had time available for other opportunities to earn money, for example by teaching. For many choristers in London, however, so-called free time was essentially captive since they did not have sufficient time to go home between sessions; if there were a morning rehearsal and a free afternoon followed by an evening performance, travel time might simply prevent them from being able to use their free time. In this respect, singers were usually worse off than their orchestral colleagues who had fewer working sessions, and this was another source of friction.

Contracts for choristers were usually for one year in the first instance and thereafter annually renewable. Some chorus agreements paid the same rate to all choristers while others paid more after intervals of so many years; one opera company paid beginners more in their first year because they needed more learning time. (All learning time for choristers was paid for and this was costly for the opera-house as the chorus all rehearsed together; new people had to be fitted in with everyone else and this led to a situation which militated against hiring young gifted singers, who might be expected to move on after a relatively short time, in those opera-houses with a long established chorus.) In practice, most choristers effectively had tenure because it seemed to be very difficult to devise a system of quality control that was acceptable to opera choruses. Arrangements existed for reauditioning singers and terminating contracts if they were found to be unsatisfactory, and there were rules about warning periods and so on but in several choruses it was basically very difficult to push people out if they do not want to go. Allied to this was the question of the age of retirement; some opera companies had retirement at 65 and others did not. In fact, in the newer opera companies or those that hired singers by the season this did not arise (or had not yet arisen) but it was an issue in the English National Opera and the Royal Opera which have long-established choruses. Not only do these factors affect the quality of performances they also have implications for the number of openings available to younger singers.

Concert and Sessions Work

Equity laid down Classical Public Concert (Concessionary) Minimum Rates (Chorus) for singers working with its approved Choral Managers. The individual's rate of pay depended upon the size of the chorus and was less for a larger group; the minimum rate gave the right to two and a half hours' performance in classical public concerts or three hours in opera. Overtime rates applied to longer performances. Rehearsals of three hours or part thereof were paid at a different rate and were in addition to the performance fee. Travel and overnight expenses had to be paid and a meals allowance. Reasonable time had to be allowed between rehearsal and performance time. If overseas travel were involved, travel time had to be paid for and singers had to be paid the equivalent of rehearsal fees during the engagement on days when there was no performance and one rehearsal fee plus performance fee when there was a performance. One day in six might be an unpaid rest day on a foreign tour. Both in the UK and abroad, radio broadcasting fees had to be arranged by negotiation and TV and gramophone recordings required an additional contract.

Sessions rates for gramophone recordings were subject to Equity's agreement with the British Phonographic Industry. The rate for a three-hour session allowed a maximum recording time of twenty minutes and that for a two-hour session, ten minutes; rates were different for classical and other gramophone recordings. There were additional payments for overtime, Sunday work, and overdubbing. Similar agreements existed for work with commercials, TV and radio jingles, and the like, for which rates of pay were higher, and contracts covered repeat fees or buy-outs.

Musicals

Equity also had contracts which related to musicals—here conditions of work were less favourable than in opera but weekly rates of pay were higher. There was a ten session week with a session lasting four and a half hours. Usually, artists in musicals did eight performances and two rehearsals a week. Conditions of employment were similar to those quoted for work in opera.

BBC

The BBC had agreements with both Equity and the ISM. Essentially (and this is a considerable over-simplification) they were similar to those for concert and sessions singers for singers doing *ad hoc* work except that they allowed for repeat fees of 50 per cent of the rehearsal and performance fee for domestic radio broadcasts after the first transmission. The ISM agreement, which was for solo performers, allowed one three-hour rehearsal on the day of the performance and the broadcast (or pre-recording for a deferred broadcast).

For BBC TV there was a minimum fee for soloists in four different categories, depending on transmission time, which covered the number of rehearsals that may be called. Chorus singers could be booked on a weekly basis, a session basis, or a combination of the two; sessions were for up to five hours' work in seven if the singer were in vision but otherwise the contract was similar to the concert singers' contract.

There were arrangements with both BBC Radio and TV for paying singers for transmission of live performances of opera and other repertoire not promoted by the BBC.

Oratorio and Church Work

The ISM had an agreement with NFMS member clubs and societies to pay soloists a recommended minimum fee, which included a same-day rehearsal along with payments for travel and provision of overnight accommodation if necessary. Church singers were not covered by any national agreement, though singers in the permanent cathedral choirs were mostly Equity members and concert and session rates and conditions applied.

Teaching Singing

The ISM had a recommended minimum fee for hourly teaching. Teachers in schools, and further and higher education were covered by national agreements applying in those sectors, though there was no standard part-time rate in universities.

Contracts Agreed by Agents

All the above contracts and conditions of work cover minimum payments and conditions with respect to rehearsals, travel, and the like. It is the job of a singer's agent to negotiate fees over and above the minimum and to make individual arrangements with respect to other conditions, otherwise singers would have no incentive to retain an agent and pay their fees, which in 1990 were usually 10 per cent of the fee for work in Britain and 15 per cent for work abroad. Apart from the performance fee and the number of performances, the singer's ability to earn money is also determined by the length of the rehearsal period included in the contract and any times and dates for which they will be not available (n/a) because of other commitments. Top singers were usually paid a fee for opera rehearsals but in the case of other soloists, the performance fee included the rehearsal period, which could be quite lengthy—usually two to three weeks for a revival but five to six weeks (or and increasingly even more) for a new production. Long rehearsal periods in which singers are not able to work elsewhere reduce the amount of work that singers can do throughout the year and hence their annual earnings. Agents make all these arrangements. Practically all established singers had an agent and many singers on the way up did too. By the time a singer was negotiating fees, he or she almost certainly had an agent.

Employment and the Demand for Singers

So far in talking about employment for singers only the total amount of work available has been discussed. What about the flow of demand, i.e. the number of jobs that are available per year? As we have seen, it is virtually impossible to answer this question because so much employment for singers is irregular. To some extent the question is answered by figures given on the number of soloists hired by opera companies, orchestras, and choral societies since that demand was the annual flow. (It should be noted that the figures given for employment included work for both British and foreign singers.) In relation to the opera choruses, however, it is possible to give figures for vacancies besides those for total employment. Altogether in 1989 there were around twenty-two chorus vacancies in the permanent opera choruses in Britain, an average turnover of 8 per cent. In the seasonal companies 123 chorus members were hired, for an average of ten weeks (though these figures are rather misleading since many of the Glyndebourne Festival Chorus of fifty-five would have continued with the tour, for which the chorus was thirty-five). Kent Opera had a chorus of thirty-six in its last season and that loss of work made a noticeable impact on the availability of work for young singers. In the case of the permanent companies operating throughout the year, the number of vacancies is usually small in relation to the total but in the case of opera companies with a shorter season, the whole chorus is hired each season. Another factor that affects the flow of demand is the policy that the opera company has with respect to the type of singer they seek for the chorus—some prefer younger singers for whom a chorus job is a stepping stone to a solo career, while other companies prefer older singers who are likely to stay with the company for a long time; demand also depends on whether or not chorus members are reauditioned and contracts terminated if they are found to be unsatisfactory; turnover is obviously higher when they are and where younger singers are preferred.

With the rest of the market for choral, concert, and sessions singers being dominated by casual employment it is anyone's guess as to the extent of demand, since demand changes almost by the hour and day.

Some observations may be made, however. In general what seems to be the case is that the more work singers do, the easier it is for them to get work. This may be because they are more experienced or because of the difficulties of making yourself known in the profession; it may also be the result of a process of natural selection, that the market works by weeding out the unfit in a way that the training process apparently fails to do. There is a huge information problem in the market for singers and searching for information is a very costly process, which the market solves by giving work simply to those who have had it before. One could put it this way: the fact that a singer has been given work by someone else is a sign that they have what it takes, that they are worth employing.

Once a singer has broken in to the magic circle he or she can put that work experience on his or her curriculum vitae and this leads to further employment. In this way an employer's information problem is solved by a former employer somewhere back down the line. Thus the market minimizes the cost of obtaining information; it is an economic solution to the problem. The result is, though, that work concentrates into the hands of relatively few singers.

But how to get started and break into the market? For those singers who mature early and are ready for professional work on leaving college, the college performances of concerts and operas provide a showcase and some, but seemingly only very few, get work immediately. This is also true of students at the National Opera Studio, which gives students considerable opportunities to get themselves known; a higher proportion, though not all of them, get work on leaving. The others have to do the rounds of auditions. All opera companies do regular audition sessions and hear hundreds of singers a year. The English National Opera in 1990 expanded its auditions as a service to young singers where they can get advice and an opinion about their potential from experienced staff. Singing competitions both in the UK and abroad also offer a platform for young singers to get themselves known, though, as we showed in the last chapter, this could be an expensive business. Few agents run auditions for young singers in Britain, preferring to hear them in performances and this is where the smaller opera companies and amateur opera companies are important, because they offer a platform for agents and other talent scouts to hear singers perform; a few agents, however, do hold auditions from time to time. (By contrast, in Germany, all agents hold regular auditions and opera-houses etc usually only hear singers on an agent's recommendation.)

But the concert and sessions world does not have the finance to audition singers and so most work in that field is obtained via word of mouth recommendation. The BBC Singers and BBC Music Department do audition extensively, hearing over three hundred singers a year—having even passed a BBC audition helps singers as it looks good on a c.v. Another experience that recommends itself highly is having been a choral scholar, particularly at Oxford or Cambridge. This is particularly the case in relation to conductors who themselves emerged from this tradition, and in relation to Early and Baroque music. Church jobs also go by word of mouth recommendation. Apart from this, any performing experience that puts singers in touch with established professionals and with the public (you really never know who might be in the audience) could result in a useful contact. The grapevine is a very cheap and an easy information network and many people are on the look-out for good singers.

The above comments suggest that economic forces affect even the way that singers are hired. There are systematic influences that determine the demand for singers, both in aggregate and at the level of the individual. These factors merit a more detailed examination.

4
Factors affecting the Demand for Singers

A number of factors affect the demand for singers. They may be divided into those factors that affect demand in aggregate, i.e. that determine the overall level of demand for classically trained singers, and those factors that influence demand for individual singers. Aggregate demand for singers is affected by the level of revenue of the opera companies, orchestras, choral societies, and the like which hire singers; this in turn is affected by the level of public subsidy to the arts and by ticket prices and sales (box office revenue). Demand for individual singers is determined by the type of work the public wishes to see (Verdi or Monteverdi), how easily one singer may be substituted for another, and the singer's voice type. Demand at all levels may also be influenced by non-market forces, such as the conditions of work and by minimum pay agreements laid down by the trades unions.

Before looking at these factors in turn, however, it is necessary to go into the theory of demand for labour services. In general, the demand for labour will be higher the lower the rate of payment, because the same number of employers or promoters are prepared to hire more hours of work (or more performances) and because more employers enter the market at lower rates of pay. The demand for labour, therefore is downward sloping. On the other hand, the supply of labour is upward sloping, with more hours of work (or more performances) being offered as rates of pay rise. Where demand and supply cross determines the rate of pay and the amount of employment. In Fig. 4.1, demand and supply cross at rate of pay P, and N is the number of hours of labour or the number of performances that are demanded and supplied.

If demand at every rate of pay increased, say because of a growth in audiences or because of increased subsidy to music or opera, a higher rate of pay and more hours of work or performances would

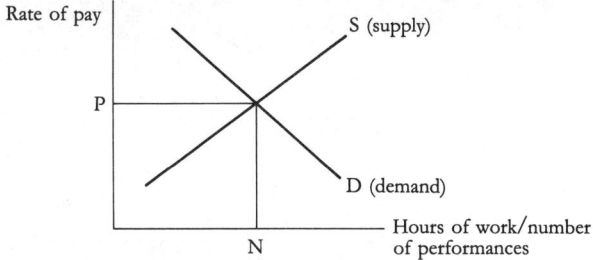

Fig. 4.1. Demand and Supply in the Market for Singers

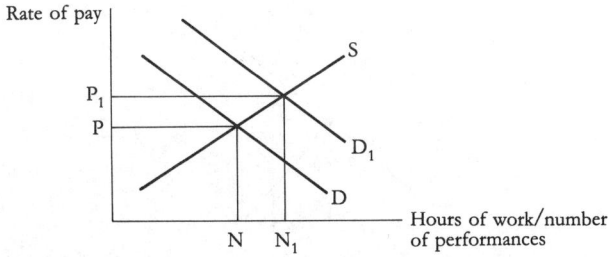

Fig. 4.2. The Effect of a Shift in Demand on Pay and Employment

be the result. This can be shown by shifting the demand schedule out to D_1 in Fig. 4.2.

Now the rate of pay would be P_1 with N_1 hours of work or performances demanded and supplied. Similarly, other factors could shift the supply schedule, for example, a change in the agreed conditions of work might change the conditions of supply such that singers would do one less performance or session a week at the same rate of pay. This would shift supply to S_1 in Fig. 4.3.

Again, the result would be a higher rate of pay P_2 but this time the hours of work or number of performances would fall to N_2 in Fig. 4.3. This is because employers' demand falls as rates of pay rise.

A third scenario is one in which, instead of the market rate of pay P being paid, one side or the other fixes the rate of payment above or below it. A minimum rate of pay might be superimposed at P_3 by trade union agreement; an example of that situation is what Equity does with its minimum rates of pay. This is illustrated in

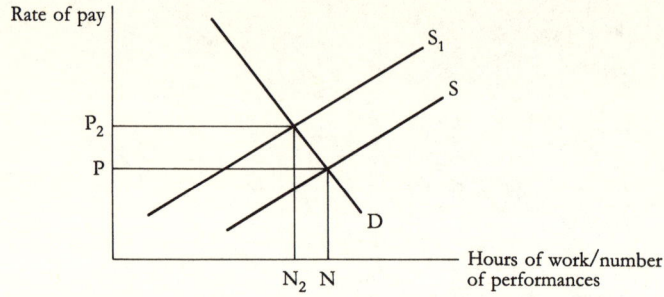

Fig. 4.3. The Effect of a Shift in Supply on Pay and Employment

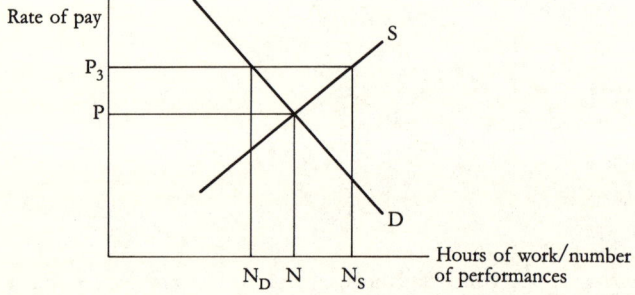

Fig. 4.4. The Effect of Imposing a Minimum Pay Floor on the Market for Singers

Fig. 4.4. In such a case demand and supply get out of balance with N_S hours–performances being offered at P_3 but only N_D being demanded by employers, i.e. there is excess supply. The imposition of a minimum rate of pay above what the market will bear causes unemployment of N_S–N_D. This shows that trade unions have to accept lower levels of employment in exchange for higher pay. Such an arrangement can only work, however, if employers agree not to hire people at lower rates of pay and trade union members do not accept work at lower rates of pay.

A fourth and final possibility is when employers combine to fix a rate of pay lower than the market rate. International opera managements attempted to do this for top-paid singers in the 1970s (see Priestley 1983: 17). In Fig. 4.5, the agreed rate of pay is P_4 and this results in an excess demand, N_D–N_S. Demand is greater than

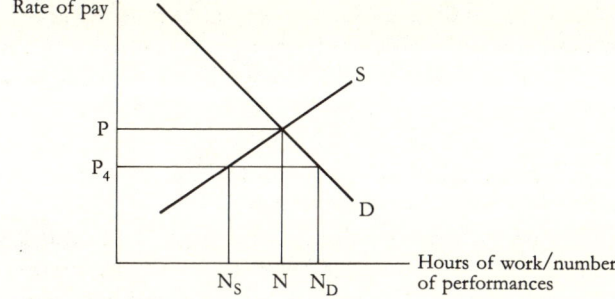

Fig. 4.5. The Effect of Imposing a Pay Ceiling on the Market for Singers

supply at P_4 because singers are unwilling to do as much work at P_4 as they would do at P and so only N_S performances were supplied rather than N. The fee-ceiling arrangement failed as operahouses broke ranks and made higher offers to singers.

The exact amount of excess supply (and unemployment) or excess demand that comes about in these situations depends (1) on how far above or below the market rate the rate of pay is fixed and (2) on the responsiveness of demand and supply to changes in the rate of pay. If demand and supply are very responsive to changes in the rate of pay, that is, demand is elastic in economic jargon, then a higher amount of excess supply or demand results and the pressure to undercut or pay over the agreed rate is the greater. It is therefore important to know what factors affect the elasticity of the demand for singers in addition to those that affect the level of demand.

Factors affecting the Elasticity of Demand for Singers

The demand for labour services is derived from the demand for the final product. In some cases the final product is the service itself, for example, a haircut or a song recital; in other cases, the final product is a good—motor cars—and the demand for labour to produce the good—car-workers—depends on the demand for the good itself. Examples of the latter case in the singing profession are to be found in opera or concert and sessions work; a record company hires singers to make records which it hopes the public will buy. In such a case the elasticity of demand for singers depends on the elasticity of demand for the record. Secondly, the elasticity of demand for singers depends on how easy it is to substitute one type of singer for another, or even to alter the number of singers required.

Where the product is a recital by one specific singer, there is little difference between the demand for the recital and the demand for the singer; even here, though, there is some distinction to be made. I may not want to hear Katia Ricciarelli sing the Wesendonck Lieder but would travel miles to hear her sing Donizetti. Taste for the singer is not quite the same as taste for the recital programme. On the other hand, exactly who is singing in a chorus or choir does not usually influence demand, but what they perform does. Thus there are two issues here, the extent of demand for the product—a concert, recital, or opera—and the degree to which substitution between performers is acceptable. Demand for singers depends on the intensity of demand for the product which in turn depends on how responsive that demand is to the price of the product—the ticket for the concert, recital, or opera—and it depends on how easy it is to substitute one singer for another. The easier it is to substitute singers, the more elastic the demand for their services. A third element in this is what proportion of the total cost of producing the final product payments to singers represent; the higher the proportion of the outlay on singers, the more sensitive the promoter becomes to the labour cost of singers, and the greater the motive to substitute less expensive (lower-paid) singers for more expensive (highly paid) singers.

These three points need to be investigated, therefore, in order to assess the elasticity of demand for singers.

The Elasticity of Demand for Concert, Recital, and Opera Tickets

Since the demand for singers' services depends on demand for the concerts, recitals, and operas that they produce, we need to know how much the public responds to changes in the price of tickets for concerts, recitals, and opera. There are not complete data on concert-going, only on attendances at a selection of London venues playing classical music (Appendix 1, Table 9) but there are complete data on opera attendances (Appendix 1, Table 10). Table 9 shows that attendances at concerts remained steady between 1984/5 and 1989/90, a period during which average ticket prices rose by 77 per cent at the Royal Festival Hall, 123 per cent at the Queen Elizabeth Hall, and 106 per cent at the Purcell Room. In opera the number of performances rose by 12 per cent and the number of seats sold rose by 19 per cent, while average prices rose from between 50 and 105 per cent (Table 10).[1] Between 1984/5 and 1989/90 the Retail Price Index rose by 28 per cent, showing that the rise in the price of concert and opera tickets was nearly double that of other prices. This evidence strongly suggests that the elasticity of demand for tickets is very low, i.e. that the public is not put off buying tickets because of their price. The same conclusion has been reached by a number of audience surveys as well.

What does this imply for the demand for singers? First, the evidence shows that the number of opera performances rose, creating more work for singers; second, it suggests that if singers' fees and wages rose, thus pushing up costs of production, managements could pass this on to audiences in the form of higher ticket prices.

Substitution

In the process of producing concerts, opera, and recitals how much substitution between singers is possible? In the economic theory of production, one type of input to the production process will be substituted for another if it is technically feasible and if relative prices change such that one input becomes relatively cheaper than another. The amount that one input, or factor of production, contributes to the total value of output is its factor productivity. In the economics of the arts, it has long been assumed that this process of input substitution to achieve greater factor productivity was not possible. This hypothesis was first put forward by Baumol and Bowen (1966) and has come to be nicknamed 'Baumol's Disease' or the 'Cost Disease' in the arts, particularly the performed arts. Baumol and Bowen argued that since the output of the performed arts is labour itself, it is not possible to introduce labour-saving changes through substituting other inputs for labour, nor is it possible to

---

[1] Strictly speaking these are not average ticket prices but average ticket yield, i.e. total box office receipts divided by the total number of paid admissions. The Arts Council computes information this way to overcome the complications of the structure of prices in theatres, as this may change from time to time. If there were only one price in a theatre then average ticket yield would unambiguously be the same as price.

raise labour productivity. Because wage rises are determined by increases in productivity in the economy as a whole, Baumol and Bowen thought that wages in the arts would rise at the same rate everywhere and hence the arts would experience steadily increasing costs. The prices of tickets to arts events would therefore rise faster than the prices of other goods and services because no productivity improvements could be made, and they therefore concluded that the arts would need increasing subsidy to survive.

There are several strands to this argument and it is not necessary to go into all the aspects of it here. In the last section it was shown that ticket prices for opera and concerts have indeed risen by more than the prices of other goods (as measured by the Retail Price Index) but this did not cause demand for them to fall. Moreover, while ticket prices were rising, subsidies to music and opera did not rise in real terms and so subsidy actually decreased as a proportion of the total income of orchestras and opera companies. At the same time these organizations, in common with other performing arts organizations, were able to make changes in response to rising costs of production, suggesting that, though there were technical constraints, substitution was possible.

Technical Constraints in Music and Opera

The type of technical constraint usually mentioned in connection with the Baumol and Bowen thesis is that *Aida* cannot be performed without an Aida and a Beethoven string quartet cannot be performed with only three players, these being examples, albeit overdone ones, of increased labour productivity. While this is patently true, there are labour-saving changes that can be, and have been, made. For one thing, *Aida* is not the only opera and others requiring fewer singers can be substituted instead if *Aida* is too expensive; furthermore, it can be performed with fifty choristers rather than a hundred. The issue, then, is whether technical constraints allow for any substitutability. Substitution could take place at two levels—the work itself and the number of performers that are needed to perform it. There is some limited possibility of flexibility in respect of the numbers needed to perform some works; there have been some notable examples of 'cut-down' versions of operas—Peter Brook's production of *Carmen* had fewer principal singers than usual and no chorus. Nearer home, the smaller opera companies in Britain—Opera 80, Travelling Opera, Pavilion Opera, Opera East, Scottish Opera Go Round—play reduced versions of operas, saving on singers and musicians. The City of Birmingham Touring Opera has a reduced version of Wagner's *Ring* Cycle, rescored for a twenty-piece orchestra. So labour-saving substitution can be made. Furthermore, there is a large repertoire of small-scale works to choose from; there are hundreds of chamber operas that could be performed, and in the orchestral field there are thousands of works that are for smaller orchestras or ensembles. But substituting different works needs audience approval (or spectacular marketing). Questions arise, therefore, about the effect of audience taste on demand.

Audience tastes are, by all accounts, notoriously narrow. Certain works will always sell well—Beethoven symphonies, Puccini operas come immediately to mind—whereas others do not. Thus orchestra and opera-house managements are limited by audience tastes if they want to sell tickets, and this constrains their ability to substitute one work for another. Audience tastes have implications for the labour costs of works; if *Carmen* or *The Magic Flute* are what opera audiences want to hear, then ten singers have to be hired to perform the principal roles in *Carmen* (four being major roles) and eighteen for those in *The Magic Flute* (with five major roles). On the face of it, *The Magic Flute* is more expensive because it requires more principals; however, it is not only the number of singers that affect the cost, it is also the type of singer—a Carmen commands a higher fee than a Pamina. Furthermore, choristers may be substituted for principals in smaller and medium-sized roles and this has become standard practice for some provincial opera companies. In choirs, soloists are often drawn from the ranks—this was the standard practice of the Monteverdi Choir. In other circumstances, an amateur choir may be substituted for a professional one, especially for repertoire that demands large choral forces. Therefore the tastes of audiences for particular works and those works' technical and artistic requirements impose constraints on managements. However, tastes do change. The swing to historically aware performance of the classical orchestral repertoire, the taste for Early music, and other such changes in taste have resulted in reductions in the number of players needed to perform even standard works. So, again, substitution has taken place and labour savings made. Another issue in relation to taste is the question what is it that audiences want from a performance—is it the orchestra, the work, the conductor, the soloist? In opera is it a particular singer, the work, the production? If audiences only want to hear Domingo as Otello no substitution is possible but usually some sort of substitution between singers, production, and the work itself is possible, and opera managements juggle one thing with another to attract audiences. Different combinations will have different cost implications and lead to different patterns of demand for singers.

Another constraint, particularly on opera managements, is the seating capacity of their theatres. If they have a smash hit they cannot move to a bigger theatre as happens in spoken theatre because of the need for an orchestra pit; therefore they cannot maximize potential revenue by selling more tickets. The fact that performance plans have to be made so long in advance—three or more years where international singers are involved—means, furthermore, that it is often very difficult to put on more performances or extend the run. On the other hand, the lack of suitable small-scale venues means that it is difficult to put on chamber operas, so the opera companies are limited by the constraints of the size of theatres which essentially restrict them to performing a certain repertoire. Some opera companies have attempted to break the constraint of seating capacity and theatre size by building new auditoria or extending their existing one. This is capital investment and usually embodies technical improvements which may be labour-saving (such as improved lighting and scene changing arrangements). However, although labour of one type may be saved, it seems that in some cases at least, more artistic labour is needed to fill the bigger stage: singers with bigger voices, more choristers and supernumaries. In fact, costs may actually rise, though revenues rise too.

A further type of technical constraint in live performance is singers' need to rest the voice between performances. Most singers singing major principal roles in opera or solo parts in oratorio or a solo recital require several days between performances. Therefore the same work with the same performers cannot be repeated night after night. This is particularly problematic in opera; if the opera company works on the repertory system different operas with different casts can be interchanged, but if the system is the stagione one (a limited season of one opera) then there can only be two or three performances a week.

FACTORS AFFECTING DEMAND FOR SINGERS

Clearly, technical constraints do exist in music and opera but there is scope, at least in the long run, for changes that save on labour costs by one kind of substitution or another. And even where the combination of different inputs is fixed, less expensive artists may be substituted for more expensive ones though there is always the risk of reducing quality; however, audiences may accept a lower quality product rather than pay higher prices. Moreover, fixed costs such as the cost of a new production can be spread by giving more performances of the same work. How much these economies really save on total costs depends upon the proportion of total costs of the production side in relation to the costs of singers and musicians, that is the labour input of performers.

Two Case Studies of Response to the Rising Costs of Singers

The only way of gaining real insight into how performing arts (and other) organizations respond to rising costs is by in-depth studies of individual organizations. Very few such studies have been undertaken and they are badly needed in the economics of the arts.

'The Choral Society Problem'

During the course of the research for this book, data were collected on four choral societies, which had put on the same oratorio (Handel's *Messiah* and Bach's B Minor Mass) repeatedly over several years. This enables a cost comparison to be made for exactly the same inputs; in the case of one choral society (B), exactly the same soloists, conductor, and orchestra were used. The data, given in Appendix 1, Table 11, are for outlays on soloists (a quartet or quintet) and orchestra; being an amateur society, the only choir costs would be for professional 'stiffening' but those costs are ignored here. I dub this the 'Choral Society Problem' because the rising costs of hiring soloists and orchestra is a perennial problem for amateur performing societies.

Choral Society B sets the scene; between 1986 and 1989 its outlay on the same soloists rose by 72 per cent but still did not exceed the outlay on the orchestra. Choral Society A, by contrast, spent almost as much on its four or five soloists as it did on the orchestra and obviously hired top singers; it had different singers each year but the same orchestra. Over the four-year period 1985/6 to 1988/9, fee payments to the singers rose by 48 per cent as compared to 33 per cent for the orchestra. Choral Society C, which had different singers and orchestras over a ten-year period, clearly changed its policy about the balance of outlay on singers and orchestra several times, as did Choral Society D. These figures are very diverse and difficult to generalize from; but they all show that the cost of putting on the same work, i.e. where no substitution of the number of soloists or the size of orchestra was possible, rose inexorably in the 1980s. But even in these four societies, we can see that the choice can be made as to whether to pay more or less for soloists and orchestra and that the societies did adapt to relative cost increases.

'Why are Covent Garden Seat Prices So High?'

This was the title of a report commissioned by the Royal Opera House from Mark Blaug (Blaug 1978). Despite its date and several subsequent studies of the financial situation of the Royal Opera House, it still represents the only published attempt to cost different policy options. One issue tackled by Blaug was what would be the effect on costs of substituting company singers for international stars? Stars cost more but their higher fees could be covered by the higher prices that could then be charged. His analysis of patterns of attendances at different-priced performances with and without stars led to the conclusion that the opera public was more sensitive to what and to whom they were listening rather than to prices; therefore audience taste played a more dominant role than changes in prices in determining demand. One aspect of taste was the work itself and Blaug estimated that for the popular works—*La Traviata*, *Carmen*, *Tosca*, Wagner's *Ring* Cycle—price rises of 10 or even more than 20 per cent would not be resisted, i.e. the demand for them was relatively inelastic. In the 1974/5 season when the study was undertaken, the highest seat price was £9.50 for two operas, *La Traviata* and *Un Ballo in Maschera*. (The cast for *Un Ballo in Maschera*, incidentally, was Domingo, Cappuccilli, and Ricciarelli with Abbado conducting and produced by Otto Schenk.) Blaug calculated that if no international artists had been used and the Royal Opera's company artists had been used instead, the savings in cost would have permitted an across-the-board reduction of seat prices of 21 per cent, but this would have meant cancelling all the higher scheduled seat prices (because there would no longer be stars) and prices would then have gone down to £5.00 or £6.50 for the front centre orchestra stalls (the calculations were made difficult by the complicated structure of seat prices throughout the house). By contrast, doing away with new productions would have reduced prices overall by 14 per cent and doing away with unpopular ballets and operas, 5 per cent. It seems reasonable to infer from this, therefore, that the Royal Opera believed that the star cast of international artists added 21 per cent to the value of the performances and that audiences were prepared to pay between £3.00 and £4.50 more (in 1975) for seats in the front stalls to hear them. Of course, this does not tell us how much more each individual artist contributed to the performances—was it Domingo or Abbado who pulled in the punters?—nor, therefore, how much higher their individual fee should be to take account of the extra contribution to the revenue of higher priced tickets. But it does indicate that the management regarded it as worthwhile to hire stars, because they added to revenue as well as to the quality of performances. This was still the case in 1990.

The Proportion of Singers' Labour Costs in Total Costs

Having established that cost-saving substitutions can be made in music and opera productions, the next question is to ask how much they would affect total costs. The answer obviously depends upon the proportion of the labour costs of singers and instrumentalists in total costs. Ideally, figures should be analysed production by production or concert by concert but such data do not exist. However, data do exist on a company basis for the four opera companies that were included in a major study of inflation in the performed arts during the 1970s by Peacock, Shoesmith, and Millner (1982). The purpose of this study was to test the Baumol and Bowen 'Cost Disease' by analysing the cost structure of a number of opera, dance, and drama companies and orchestras; the proportion of cost increases during the 1970s attributable to labour were then calculated in order to see whether labour costs had been the cause of cost inflation.

The procedure adopted in the study was to examine the total costs of each organization in 1980/1 and to work out the proportion of total costs that were expended on each budget item. These were then used as weights in the calculation of the importance of each item in contributing to cost inflation during the ten year period studied from 1970/1 to 1980/1. The relevant figures are those in the first column in Appendix 1, Table 12 for the English National Opera, Scottish Opera, and Welsh National Opera (unfortunately, it is not possible to interpret the figures for the Royal Opera House in the same way). The only budget items given in Table 12 are those relating to singers and musicians, though, of course, every type of input was included in the published study.

In 1981 company principals and guest artist singers comprised 7 per cent of the total expenditure of the English National Opera with the chorus, actors, and children also comprising a further 7 per cent of total expenditure. Thus at the English National Opera, 14 per cent (at most) of its total costs went on singers. Comparable figures for Scottish Opera were nearly 13 per cent for principal singers and under 8 per cent for the chorus, a total of 20 per cent; at Welsh National Opera, 6 per cent of total expenditures was on guest artists including conductors, and 11 per cent on the chorus, dancers, and actors, i.e. less than 17 per cent of total costs was spent on singers.

These figures are surprisingly low and, the study being now over ten years out of date, it might be said that things have changed. As part of the research for this book, the main opera companies were asked what proportion of their total budget in 1988/9 went on payments to singers. The answers ranged from 12 to 36 per cent and so were in more or less the same region as in those reported in the Peacock, Shoesmith, and Millner study. Where individual comparisons with the earlier figures were possible, the percentages seemed to have risen; however, it is not certain that like is compared with like. Even so, the proportion of the cost of singers in total costs of opera companies is relatively low.

Table 12 also shows the extent to which singers' fees and wages rose during the inflation of the 1970s and this can be related to the index of aggregate costs (total costs) calculated for the four opera companies (now including the Royal Opera House) in the Peacock, Shoesmith, and Millner study. During the period 1970/1 to 1980/1, the Aggregate Cost Index of the Royal Opera House almost quadrupled, (rising from 100 to 396, a rise of 296 per cent). The cost of the chorus rose by 299 per cent and thus remained a constant proportion of total costs. At the English National Opera, the Aggregate Cost Index rose by 265 per cent; the cost of the chorus, actors, and children rose by 179 per cent, indicating that their contribution to total costs fell; the cost of company principals and guest artists rose by 266 per cent, i.e. at the same rate as the increase in Aggregate Cost Index, and so remained a constant proportion of total costs. At both Scottish Opera and Welsh National Opera, payments to singers, principals and chorus, increased more than their Aggregate Cost Indices, showing that the cost of singers rose proportionately from 1973/4 for Scottish Opera and for Welsh National Opera from 1975/6 to 1980/1, the year of the study.

Thus over the decade of the 1970s increases in payments to singers (but not necessarily singers' pay; there may simply have been more singers) seem to have caused increases in total costs at the then newer opera companies, Scottish Opera and Welsh National Opera, but not at the older established companies, the Royal Opera and English National Opera. There is reason to believe that the older companies, being in a steadier artistic and financial state represent the typical relationship of singers' costs to total costs. Without repeating this valuable exercise for the 1980s this can only be speculation. One cannot say categorically, therefore, whether payments to singers constituted a constant or an increasing proportion of total costs but, if the Royal Opera and English National Opera were the yardstick, then they were a constant proportion.

The Peacock, Shoesmith, and Millner study found that the annual average increase in the total costs of all the opera companies during the 1970s was 13.75 per cent. During the same period the Retail Price Index rose by 13.5 per cent. Comparable figures for opera, dance, music, and drama are to be found in Appendix 1, Table 13. Cost inflation in opera and music, therefore, was exactly the same as that of all goods and services in the economy. This is a refutation of Baumol's Disease in these two areas of the performed arts, at least. Why was this? One reason, perhaps the main reason, is that payments to singers at the Royal Opera House and the English National Opera, though they kept price with inflation as measured by the Retail Price Index, did not increase at the rate of increases in wages elsewhere in the economy, as measured by the Average Earnings Index. This is shown in Appendix 1, Table 12. Take, for example, company principals and guest artists at the English National Opera: their pay rose over the period at the rate of the Retail Price Index but less than the Average Earnings Index. Baumol's hypothesis (Baumol and Bowen 1966) was that labour costs in the arts would be pushed up by wage increases in other occupations; the evidence of the Peacock, Shoesmith, and Millner study showed that, on the contrary, the wage increases of performing artists were in general lower than those in other sectors of the economy. Incidentally, the study showed that payments to administrative staff, cleaners, etc. did rise at the rate of other wage increases. In other words, the pay of people who could work outside the arts sector keep up with that of other workers; the pay of artists 'trapped' in the arts did not.

This completes the analysis of factors affecting the elasticity of demand for singers' services, i.e. how responsive employers are to changes in singers' fees or wages. It has been argued that, in a situation in which the public has to have Singer X and will pay any ticket price to hear him or her, no substitution is possible and so the elasticity of demand for the singers' services is very low. But in most situations, substitution between singers can be made; managements can reduce their costs of production by altering the number and type of singer they employ. How strong the incentive to do this is depends on the proportion of total costs that is spent on singers; this seems to be relatively low, in British opera companies at least. In an international opera-house, such as the Royal Opera House, commitment to high quality international singers complicates the situation because ticket prices can be raised to finance their higher fees; what is the artistic contribution of the individual singer to high quality products is a complex issue, but even so we can ask what considerations go into the economic decisions of the opera company. Much of the above analysis hinges on the public's responsiveness to ticket price changes and on their tastes. Box office revenue from ticket sales has become increasingly important in the finance of music and opera in Britain in the 1980s and this, together with the level of public subsidy, affects the level of demand for singers.

Factors affecting the Level of Demand for Singers: Subsidy to Opera and Music

In most European countries the level of demand for singers is strongly influenced by the level of public subsidy to opera and music. In Britain in 1990 the Arts Council was the body responsible for distributing central government subsidy to the performing arts; it directly subsidized arts organizations in England and distributed funds to the Arts Councils in Wales, Scotland and Northern Ireland, and to the Regional Arts Associations, with which they in turn subsidized organizations in their areas. In 1989/90, the Arts Councils' subsidy to opera was £26.8 million and to music £24 million (see Appendix 1, Table 14). Table 14 gives details of subsidy from the Arts Councils to music and opera from 1983/4 to 1989/90. (The figures for music are not always separately reported and are anyway complicated by the abolition of the Greater London Council.) In addition to central government subsidy local authorities also spent £6.7 million on music and opera in 1987/8. There was a small but steady increase in the total subsidy to opera from 1983/4 to 1989/90, but the increase of 16 per cent over the period was swamped by inflation, which was double that, the net result being a fall in the real value of opera subsidy.

How important was subsidy to the opera companies? In 1983/4 63 per cent of the principal opera companies' income was from public subsidy; by 1989/90 this had fallen to 49 per cent. Box office and earned income rose from 31 to 38 per cent, and sponsorship and donations increased from 6 to 13 per cent. Subsidy was (and is) therefore very important to opera and even though it fell in real terms during the 1980s, it still comprised half the opera companies' real income. And although subsidy fell, box office earnings rose because prices and the number of opera performances rose (shown in Appendix 1, Table 10). It cannot be said, therefore, that the real decline in subsidy to opera reduced the demand for singers, because the number of performances actually rose but it may have caused changes, such as the type of substitution discussed earlier, that affected the pattern of demand for singers.

The extent to which opera companies could sustain further cuts in the real value of subsidy without having to cut the number of performances depends on the willingness of audiences to absorb ticket increases. Thus the argument comes full circle, back to the relationship between the elasticity of demand for tickets and the elasticity of demand for singers' services. Before leaving this point, it is interesting to ask what would be the effect of a significant rise in subsidy to opera? Would it result in more performances, lower prices for opera-goers or simply increased payments to singers? As there have been no studies on how performing arts organizations use subsidy, a prediction cannot be made. It was always said that the sudden, huge increases in arts subsidy in France in 1981 merely resulted in higher payments to artists and arts administrators, at least in the short run, rather than more output of the arts.

The Response of Opera Companies to the Fall in the Real Value of Subsidy

Opera companies have responded in two ways to the fall in the real value of public subsidy during the 1980s—by substituting revivals of old productions of operas for new productions and by co-productions shared between opera companies. Both are intended to spread the fixed costs of productions and to enable the reduction of average costs via economies of scale. Both have implications for the pattern of demand for singers.

The fixed cost of an opera production consists of the weeks of rehearsal time prior to performance; the fees of the producer, designers, etc.; and the cost of sets and costumes. As more emphasis has come to be placed on the production, i.e. on visual and dramatic presentation of operas, pre-production rehearsal periods have lengthened and the outlay on producers' and designers' fees and sets and costumes has increased. This is the case with musicals as well as with opera. The fixed costs to singers are the time spent in learning the opera. The learning and rehearsal time of the chorus is already paid for as part of the chorus contract and therefore need not cost the opera company more; however, learning time for the chorus constitutes work and is subject to agreed conditions of work (the number of sessions in a week and restrictions about the timing of rehearsals in relation to performances). If learning and rehearsal time cannot be dovetailed with performances, managements may have to pay overtime. This is one way in which the Equity conditions of work outlined in the last chapter affect the demand for singers. When the opera company has a large number of performances a year, as is the case with the English National Opera, they effectively have two choruses (one chorus working in two groups) in order to deal with these problems.

The fixed cost of learning and rehearsal time of principals who are contracted for a number of performances of an opera is free to the opera company, being borne entirely by the singer. It makes no difference to the opera company whether the initial rehearsal period is four or eight weeks as far as principals are concerned because it generally does not have to pay for their rehearsal period.

But to the singer the longer the rehearsal period the lower the real value of the performance fee, averaged out over the time involved. A £1,000 performance fee for six performances may seem a lot of money. But if the £6,000 is for a rehearsal period of eight weeks followed by four weeks' performances, it represents the income of three months' work. The singer may be able to interweave rehearsals for one work with performances of another but that may be difficult if work is in different countries. The more performances of the same role in the same production that a singer can get, the more fixed costs are spread and the higher their earnings per period of time.

When an opera production is revived, rehearsal time is much shorter than for a new production. British opera companies usually have three or four weeks to rehearse a revival (by contrast, in Germany and elsewhere, revivals are frequently put on at extremely short notice, sometimes without any further rehearsal: See Appendix 3). If the same principals are engaged for the revival as were engaged in the original production, this benefits both singers and management because they both reap economies of scale. Accordingly, there is a strong financial motive on each side to hire the same singers for revivals. From the point of view of the market for singers, this tends to concentrate the demand on to a few singers who get a great deal of work while others get relatively little. The more productions of different operas there are, the more work will be distributed between a number of singers, because different operas demand different voice types. Since revivals have to be popular operas or tickets would not sell well, demand not only concentrates on a few singers but on a few familiar roles and this allows singers to make a career by specializing in a few roles.

Co-productions have the same effect but may now internationalize the process, since many co-productions are in partnership with foreign opera companies. They are sometimes linked to recordings as well. The effect of these reactions on the part of opera companies to the falling real value of public subsidy, therefore, is to focus demand on fewer singers, thus widening the gap between the successful and the unsuccessful. Nor is this only a British phenomenon. Public subsidy to the arts is under threat in many other European countries and opera companies have reacted in those places too. Italy, with its stagione system of opera production, now has fewer opera companies and the main opera companies put on fewer productions and performances than formerly. Thus the overall level of demand for singers depends on the level of public subsidy and even where the number of performances does not fall when the real value of subsidy falls, as has been the case in Britain, the pattern of demand for singers may be affected.

Conclusion

In this chapter several factors affecting the demand for singers have been discussed. The derived nature of the demand for singers has been shown to influence both the aggregate level of demand and also demand for different types of singers. Ultimately, therefore the demand for singers depends on the demand for the works that they perform in. This in turn depends on public taste and on ticket prices. So far, at least, the demand for tickets in Britain seems not to have responded very much to increases in ticket prices and in opera, the only art form employing singers for which there are data, the number of performances has risen during the 1980s. In both opera and music, managements have had to rely increasingly on box office revenue rather than public subsidy but in opera that still constitutes half the main opera companies combined revenues (the exception to this is the Glyndebourne Festival Opera, which is privately financed). Big changes in public subsidy could, therefore, affect the demand for singers. The opera companies' policies in relation to these changes have altered the pattern of demand for singers.

Increases in ticket prices in music and opera have undoubtedly been made in response to rising labour costs of singers and musicians but these costs do not constitute a very large proportion of total production costs in opera. Payments to singers and musicians have not increased at the rate of wage inflation in the economy, though they kept pace with the cost of living in the 1970s. Baumol and Bowen's predictions have not been confirmed in British experience for this reason. In addition, there is more flexibility in the performing arts than they supposed. Substitutions of all kinds have been made in response to changing costs and this has meant that labour costs (and hence singers' earnings) have not risen as fast as they otherwise might have done.

If we could understand the pattern of demand for singers better, it would help to answer the question as to whether or not there is an oversupply (or excess supply over demand) of singers. It has been shown that demand is not rigid; substitution of less expensive singers does take place, creating openings for younger and/or less gifted singers. However, there is also a tendency for the market to concentrate on a few stars. This trend has considerably affected the distribution of singers' earnings.

Part III
The Returns to Training

5
The Earnings of Singers

Earnings are determined by the amount of work that is undertaken and the rate at which that work is paid. In any labour-market the rate of payment—the wage rate or fee—is determined by the interaction of the forces of supply and demand. Even where the wage rate is fixed by collective bargaining, these forces have to be taken into account. The wage rate per hour or the fee per performance is both the basis of income to the supplier (the singer) and a cost to the employer. The individual singer decides how much work he or she is willing to undertake at various rates of pay, but how much work he or she actually does, and hence how much is earned, depends on conditions in the market. Most singers and their agents understand this very well and consciously set their fees according to 'what the market will bear'.

It is obvious from the discussion of employment in Chapter 3 that the earnings of singers are likely to be very difficult to measure. Only a relatively small proportion of singers are salaried, the rest working freelance with enormous variation in the fees paid per performance and in the number of performances that a singer does in a year. As a result there is considerable variation in annual earnings between individual singers. An individual singer's earnings vary from year to year and, furthermore, may vary with age; older singers may find it increasingly difficult to get work or may give up performing and move into teaching; young singers earn little at the beginning of their careers, partly because they are paid lower fees and partly because they get less work. Even singers who are on annual contracts experience some variation in their annual earnings despite having a regular weekly salary, because the amount of overtime and other extra payments, such as income from recording, vary.

How is one to find out about singers' earnings? The best way might seem to be a survey of singers to find out what they earn. This is not an easy thing to do, however, even if one had the resources to mount a full survey. The precondition for doing a survey is having a complete list of the population that is to be surveyed so that a sample of people to be included in the survey can be drawn up; then individual names can be drawn from the list in such a way as to ensure that bias is not introduced that would make the results unrepresentative; for example, if women earn less than men, a sample survey which contacted proportionately more women than men would introduce a downward bias into the results. But, membership of the singing profession is not easy to define. One way of getting such a list is to use membership of a professional association or trade union but the problem is that all singers may not be members. In Britain there is no association or union exclusively for singers; singers belong to the ISM (solo performers and teachers) and to Equity (those working on stage and in TV, films, and recording work). Many belong to both. Each of these organizations had undertaken earnings surveys of their members just before research for this book began; however, the results related to the full spectrum of membership, rather than just to singers, and it was not possible to extract information about singers alone.

Because of the difficulties of conducting a random sample survey of singers, and because it seemed unlikely that those concerned would co-operate with another survey so soon, I took another approach in the research for this book. I interviewed a selection of singers and agents about fees, likely numbers of performances, what constituted a full-time workload, and earnings; in addition, employers were asked about what fees and salaries they paid in 1988/9 and a sample of fee payments made by amateur performing societies was analysed. Information was also collected on negotiated minimum rates of pay from Equity and the ISM. From these sources it has been possible to draw up a comprehensive picture of fees paid during 1988/9 and also a good picture of the workload of singers at different stages and at different levels of the profession. The singers were carefully selected according to several criteria—being at different stages of their careers, having different voice types, working in different branches of the profession and so on. The survey of singing-teachers undertaken for this book, which was discussed in Chapter 1, is also relevant here because it contained information on earnings. Thus a fair amount of data have been assembled from these various sources and results can be cross-checked.

Besides gathering information about singers' earnings, it is also important to try to explain why the labour-market for singers works as it does. In this chapter, the evidence is presented and the question is raised what this indicates about the workings of the market for singers. In the next chapter, armed with the information about the earnings of singers in Britain, the question of whether it pays to become a singer is considered; given the investment that must be made in training as a singer, can lifetime earnings be expected to recoup the cost of that investment? This issue is relevant both at the individual level to young people making a career decision and to educational policy-making. Ultimately, that is what determines the supply of singers and it should also influence decisions about how many singers are trained.

Surveys of Singers' Earnings in Britain

The two surveys that included information on singers' earnings in Britain were the survey of Equity members and the survey by the ISM. The survey of Equity members undertaken by Dr Barry King was a one in twelve random sample of Equity membership between April 1987 and March 1988, structured for sex and age (King 1989); Equity membership includes a wide range of performing artists, from actors (the most numerous) to ice-skaters, and singers accounted for only 7 per cent of the respondents to the survey. In a situation in which the overall response rate was fairly low even for a survey of its type (19.4 per cent) this means that it is very hard to have much confidence in the results of a small sub-group such as singers; moreover, singers in Equity include people working in musicals, variety and so on, who have not necessarily had a classical training. It is only possible to be sure that singers identified in the survey as working in opera had had a classical training, and they constituted under 5 per cent of the sample. In fact, there is reason to believe that singers as a group did better than the Equity membership as a whole, since those identified as working in opera worked three times as many weeks as the respondents' average (King 1989: 10). Therefore, the overall results of the Equity survey almost certainly underestimated the earnings and employment of singers.

The results for all types of performers showed first that 11 per cent of the sample had been unemployed for the whole year, and that nearly half of those who did work worked for fifteen weeks or less in the year; and secondly that median gross earnings from professional live performance were £3,250 (or £3,750 including repeat fees, use fees, royalties, etc.) in 1987/8.[1]

The ISM undertook its survey in 1989 of all 850 members of its Solo Performers Section, which included 230 singers. The results were analysed and published by Feist and Hutchison (1989: vol. iii); as they point out, many of this Section's members were the older and more established musicians and 'by definition it was likely to include a higher proportion of financially successful solo artists than any general cross section of the music profession' (p. 10). Although singers were rather underrepresented in the sample (they constituted 27 per cent of the ISM membership but only 19 per cent of the sample), the survey was probably representative of the singing profession as a whole. It is true that relatively few of the top opera singers were members of the ISM; on the other hand, because it was a survey of solo performers, it did not include the hosts of singers on the concert and sessions scene who are choral singers rather than soloists. The results showed that just over one-third of members earned less than £10,000 and another 30 per cent earned between £10,000 and £15,000; of the remaining third, half earned between £15,000 and £20,000 and the other half between £20,000 and £50,000; 4 per cent earned over £50,000. The median income, therefore was just over £12,000. These figures related to 1987/8, the same year as the Equity survey, and showed that ISM members were earning over three times the earnings of the typical Equity member.

The above figures related to gross income and so considerably overestimated pre-tax income net of payments for subsistence, travel, self-employed National Insurance contributions, etc. My enquiries from singers and agents have shown that typically 35–40 per cent of gross income is accounted for by this type of expense (see Appendix 1, Table 18). This brings the ISM median figure for pre-tax net income down to about £8,000 for 1987/8 and it is that which

---

[1] The term median crops up a number of times in this chapter. Where earnings (or any other data, such as days worked, or age) are not evenly distributed in such a way that the average, i.e. the arithmetic mean, cuts the distribution in half, the median is a better measure of the typical figure. The median is the statistic that finds the mid-point (i.e. the 50 per cent mark) of the distribution in question; thus median earnings are a better measure of typical earnings than average earnings, in a situation in which a few people earn a great deal and many people earn considerably less, when the high earners would pull up the average figure making it unrepresentative.

should be compared to salaries in other professions or occupations. Assuming that musicians' earnings rose at the rate of inflation, the equivalent figure for 1989/90 would be around £9,300. In 1989 average non-manual earnings for men were £19,500 a year and £10,800 for women (see Department of Employment 1989) for the age range of 40–9, which seems the appropriate one to use for people who were well established professionally, as the solo performers of ISM were.[2] Thus neither women nor men in ISM were earning an income that was comparable to the average for the economy as a whole, according to the results of this survey.

Only limited reliance can be placed on the results of these surveys because response rates were low. Low response rates introduce an unknown bias into results—for example, if the only people with the time to fill out the questionnaire were those who are unemployed, the survey would produce a disproportionate number of unemployed people in the results. In fairness, though, it has to be said that all such questionnaire surveys have a tendency to get low response rates. But an equally serious problem was that the surveys covered all types of performing artists and musicians, not just singers, and so were only relevant if it could be assumed that singers were typical of the membership of these organizations as a whole.

The Earnings of Salaried Singers

The only groups of singers in receipt of an annual salary while this book was researched between 1988 and 1990 were singers in the choruses of the permanent opera companies, those working in musicals, the BBC Singers, professional singers in cathedral choirs, and company singers in the permanent opera companies. The latter were chiefly at English National Opera and the Royal Opera House, though the other companies had one or two company singers apiece. Other opera singers, principals and chorus, in the seasonal opera companies were also on salaries for part of the year. The order in which this information is presented below is deliberate in that it starts with singers for whom a salary was mostly the sole source of

---

[2] It is a serious weakness of the ISM survey that it did not collect age-specific earnings or collect information about the age of its members. This is important because earnings differ considerably with age and signify rather different things at different ages; for example, a young singer earning £6,000 is not doing badly but a singer in mid-career would be doing badly with a £6,000/year income.

earnings and moves through to those for whom the salary represented a decreasing proportion of their total earnings.

The Permanent Opera Choruses

One sub-section of the singing profession for which there was reliable information on earnings was singers working on a full-time annual contract with one of the permanent opera company choruses in Britain. These singers were on an annual salary and in addition received extra payments for a number of different items, especially overtime (details were given in Chapter 3). Salaries differed somewhat between the five companies and payment differed according to experience within the companies (except the English National Opera chorus, in which all choristers were paid the same). In 1988/9 weekly basic rates of pay varied from £163.25 to £205. This meant a basic annual salary of £9,000–£10,660 a year. Most choristers, however, were being paid near the top of the scale, typically over £180 a week. Extra payments for overtime, recording, etc. and a (then) non-taxable vocal maintenance allowance added on average 20 per cent to this, so most choristers were earning somewhere between £10,800 and £12,800 a year before tax, with employers paying contributions to National Insurance and to a special pension plan for opera singers. Comparison with the results of the last section therefore shows that choristers in the permanent opera choruses fared better than the average solo performer in the ISM survey and very considerably better than the average Equity member. Moreover, for women this clearly compared well with average earnings in non-manual work, though for men it was still much lower than the average in other non-manual occupations (see Appendix 1, Table 7).

Musicals

Singers in musicals were mostly on weekly salaries. The lowest rate that was paid in the West End of London in musicals was £250 in 1989 for the chorus, with principals commanding £450 a week upwards; a middle-range principal part was paid around £600–£800 a week with £1,000 plus for a big name. Top stars were paid £8,000–£9,000 a week and really top rates were reputed to be anything from £10,000 to £24,000 a week, with the stars often getting a share of box office profits as well. So in musicals in London chorus singers earned a minimum of £13,000 in 1989, principals typically between £31,200 and £41,600 a year, with £52,000 plus for lead players. (These figures were based on performing weeks; rehearsals were paid at the Standard Equity rate.) Outside London, earnings were considerably lower, probably around £200 a week for a 'name' singer (£10,400 a year for a full year) or £180 a week for chorus. However, musicals are notoriously unstable; once they are established employment in them could be regular for a few years, though many fail after a few weeks or months. And singers work more sessions in musicals than in opera (see Chapter 3).

BBC Singers

In 1989 the minimum basic pay of the BBC Singers was £13,500 with small percentage increments for years of service going up to an extra 10 per cent for ten years' service and with employers' National Insurance contributions paid, but no pension. There was also the possibility of some overtime payments.

Cathedral Choirs

I was unable to get comprehensive information about the salaries of professional singers in cathedral choirs. The payment situation was somewhat complicated by payments in kind, such as free accommodation, or the linking of choir work with a teaching post in the choir school which was a common feature of cathedral work for men. One big London cathedral paid a basic salary of £6,000 in 1988/9 with a good pension scheme and employers' National Insurance contributions paid in addition. Extra earnings were about £1,500 in the cathedral and choristers had quite a lot of time free in which to work elsewhere, for example, in concert and sessions work.

Company Artists in the Permanent Opera Companies

Salaries of company artists in opera varied from £180 to £600 a week in 1989 depending on experience; this latter figure is misleading, however, because top artists might also be paid a performance fee as well. Young principals were being paid somewhere between £180 and £275 a week, £9,360–£14,300 per annum, with employers' contributions to National Insurance being paid. This was not considerably less than the average non-manual weekly wage in 1989 of £277 for men aged 25–9 and for women it compared well with the equivalent figure of £206 a week (Department of Employment 1989). It has to be recognized, though, that these young principals were already on the road to success by virtue of having such a position with an opera company. The great difficulty with getting information about earnings of singers in general is that it is easier to find out about the successful rather than the unsuccessful who, almost by definition, are hard to identify.

Seasonal Opera Companies

The difficulty of estimating the earnings of singers who work with seasonal opera companies is that although it is possible to find out what they can typically earn with one company, one does not know what, if anything, they earn the rest of the time. This is obviously the case with freelance singers but it is also true of singers on a salary with seasonal opera companies. Singers with Opera 80, Glyndebourne Festival Opera Chorus or Glyndebourne Touring Opera chorus, Travelling Opera, Pavilion Opera, Opera Factory–London Sinfonietta, and British Youth Opera were paid at or just above the Equity Minimum Rate under the National Agreement for Opera Singers, which was £165 a week in 1990 with a Touring Allowance of £149.60 a week. Some companies paid a small performance fee on top of that to soloists and/or made extra payments for travel time (on top of travel costs, that is). Thus a singer doing, say, a fifteen-week contract with one of these companies would have earned a basic salary of £2,475, as well as the Touring Allowance when touring, in 1990. The Touring Allowance was regarded as being more or less adequate in ordinary small hotels or guest houses, but not enough in big cities or for slightly superior accommodation; some singers did, in fact, manage to save on their Touring Allowance but others ended up paying out of their own pockets for better accommodation. (Most singers now have to pay tax on the Touring Allowance under Schedule E.) If a young singer could fit in two or more of such contracts a year they could earn enough to keep body and soul together. This was possible with the two Glyndebourne choruses, because they ran consecutively, but other opera companies' performing seasons overlapped, and so a singer could not fit in work with more than one company.

The Earnings of Freelance Singers

The majority of classically trained singers worked freelance and the only information about earnings that could be gained systematically was fee payments. Fees are, of course, only half the story; the other half is how much work at those fees singers were able to do. Later in the chapter, an attempt is made to relate fee payments to annual earnings from data supplied anonymously by singers' agents and from survey data for a selection of singers. First, though, details are given of fees in concert and sessions work, opera, oratorio, and orchestral concerts.

Concert and Sessions Work

This consisted of live performance with choirs and ensembles of various sizes including radio and TV programmes with an audience, such as 'Friday Night is Music Night'. Extra chorus work with opera companies or 'stiffening' for amateur choirs was also paid at these rates. Recording work (records, TV, film, etc.) was paid at a higher rate. Practically all concert and sessions work was paid at the relevant Equity Minimum rate; since there are many of these it is impossible to summarize them all. A selection is offered here. In mid-1989 Equity Minimum rates were as follows:

Classical Concerts (for a 3-hour performance session of opera or 2.5-hour session in a concert):
    Choruses of: 9–15 singers   £35 per singer
                 16–19 singers   £31 per singer
                 20+ singers   £28 per singer
(plus £23 per 3-hour session for rehearsals).

Classical Recordings:

Choruses of:		Session	Fee
up to 20 singers		3-hour session	£50.80
		2-hour session	£40.30
over 20 singers		3-hour session	£48.45
		2-hour session	£36.30

(plus £35.40 for a 3-hour (or part thereof) rehearsal session).

Session Singers (all recording other than classical):
    3-hour session   £65
    2-hour session   £50.95.

(Note that conditions of work with respect to these Equity agreements were given in detail in Chapter 3.)

Commercials and TV jingles and film work:
    1-hour session (commercials)   £60
    1-hour session (films)   up to £120.

For all types of work Equity's Annual Report lists current rates.[3]

The rates for recordings and sessions usually included a buy-out of the use rights and royalties. However, for jingles and commercials the singer got repeat fees which could add up quite considerably. It has to be borne in mind that with repeat fees, etc. a singer could still be earning after retirement. (This is true of dead singers as well, whose recorded work may continue to produce royalties for their estate.)

It is not easy to generalize about what singers could earn from this type of work. The only guide I found was hearsay—not, perhaps, a very scientific approach but the only one available short of a proper survey of this branch of the singing profession. Hearsay came from singers themselves, fixers (people who hire singers for the job in hand) of choirs and choruses, and from other people connected with the business. In 1989 a very good young concert and sessions singer could expect to earn between £15,000 and £20,000, going up to £30,000 to £40,000 for experienced 35-40 year-olds. Experienced singers would probably have specialized in a particular field and made all possible contacts with fixers and promoters. That could have its negative side; if the fixer retired or became less active or popular, the singers working regularly with him or her would suffer a reduction in earnings. It seemed to be the case that concert and sessions singers found it increasingly difficult to maintain the same amount of work as they got older.

Solo Performance Fees

In general, solo performance fees may be expected to vary as between the different branches of the singing profession and also within a particular field for good economic reasons. Concerts with orchestra tend to pay a higher rate than opera, because concerts are usually one-off events and so the fixed cost of studying the part cannot be spread over several performances; in opera the singer is hired for a number of performances and can spread these costs. But even within the field of orchestral concerts, fees vary considerably, with very high fees being paid by top orchestras in Britain to top

---

[3] It is not due to laziness on the author's part that figures, some of which could be updated, have been left at their 1989 or 1990 values. It has purposely been done so that comparisons can be made at one point in time between singers' earnings in different types of work.

singers and low fees being paid by smaller or semi-professional orchestras. We expect this to be a feature of performance fees for solo work in opera, oratorio, concerts, and recitals. Singers have a concept of 'my fee' but they are prepared to bargain (or, to be more accurate their agents bargain on their behalf); if the promoter really wants a particular singer he or she must pay what the singer demands; if the singer wants to do the work offered at a lower fee he or she may accept it, but there must be some good reason, such as the opportunity to sing a particular work with a particular opera-house or conductor—and so the bargain is struck. As stated at the beginning of the chapter, the concept of 'what the market will bear' and bargaining about fees is very well understood by both sides. This has always been so since singers started to negotiate for their pay (see Rosselli 1991: 82–7). Some agents actually issue 'price lists' of the usual fees of the soloists they represent; these take the form of a fee range within which the singer offers her or his services and the agent operates within that range, expecting the higher end of the fee range for a one-off concert with a big orchestra but accepting the lower end of the fee range upon occasion. Most singers and all singers' agents have a clear concept of the going rate for a particular type of work for a singer of a particular type of work for a singer of a particular standing, even if they do not publish it.

Singers are essentially ranked by their fee and the fee is a signal to the market of the singer's standing. Many singers would rather sing for nothing, say for a charity performance, than reduce their fee because doing so would give the wrong signal to the market. A singer who is rising in popularity will expect her or his fee to rise and, as singers get bookings quite far in advance (two or three years is normal for sought-after singers), future fee rises have to be anticipated. Singers who are waning in popularity will play safe by not raising their fee. However, if a singer is not getting enough work, lowering the performance fee is unlikely to attract more dates; indeed it could easily have the opposite effect because the market would read this as a signal that the singer is fading in popularity. In a situation in which the quality of a singer is difficult to measure and really only fully understood by relatively few people, the price of a singer's service, the performance fee, tends to be used as an index of quality or talent. Because of this, the labour-market for singers works differently from other markets in the economy in which there is some objective measure of the output of the services of labour.

This is an important aspect of the economics of the singing profession to which we return later in discussing the distribution of singers' earnings.

What, then, were fees for solo singers in Britain in 1989? The answer was anything from zero to £8,000! These were figures that emerged from my enquiries and observations of various individuals and organizations. It is convenient to give full details by the type of work involved.

Opera

Information about soloists' fees in opera comes from several sources—interviews with singers, agents, and the permanent opera companies in Britain and a postal questionnaire to the smaller opera companies.[4] Opera companies were asked to give their average or typical fee paid in 1988/9 and also the range of highest to lowest fee. The lowest fee, strictly speaking was zero, since even professional singers would work for nothing in a semi-professional opera company if they had the opportunity to perform a major role and gain experience performing it. The major British opera companies, however, all paid a performance fee for any solo line, even when it was performed by a singer from the chorus; such fees were sometimes a token in amount (£7 or £10, for example), but more could be earned from the overtime payments associated with such work. What differed as between the companies was the proportion of the smaller solo roles in any particular opera that got assigned to choristers as opposed to principals. Here policy varied; at the Royal Opera even quite small roles were usually filled by principals, possibly from their own company or by their Young Singers. By contrast it was the policy of provincial British opera companies to have choristers or their one or two young company singers fill not only small roles but also what might be called medium-sized roles as well: for example, in the *Marriage of Figaro*, only the leading roles (Count, Countess, Susanna, and Figaro) would be done by principals in at least one of the permanent British provincial companies, with

[4] The Royal Opera management co-operated generously with this study but would not reveal soloists' fees. This is understandable since one would expect their fees, being internationally determined, to come out top in any data that was given by fee range. However, the Royal Opera is not the only source of information about its fees; in some cases singers or agents have quoted fees which are relevant to the Royal Opera. It should be made clear that, where this occurs, the source of such information was not the Royal Opera itself.

choristers taking the other roles. Another category of work in relation to which policy varied between companies was whether or not they hired outside covers (understudies);[5] again, covering smaller parts was often assigned to choristers or company singers but it might also be given to an outsider. In 1988/9 covers for most roles were paid £135 a week during the rehearsal period and £30–£50 to stand by. If they went on they were paid anything from £175 to £700 depending upon the role and who the cover was. (If a singer in a major role at the Royal Opera fell ill, efforts were usually made to find a suitable international singer as a substitute; this was not the same as covering and the singer would be paid the usual performance fee.)

Coming back to the main point then, what was meant by a principal soloist's fee differed in different companies. That apart, fees in major British opera companies in 1988/9 averaged about £250 for smaller parts and were typically £500 for principal roles. The range of fees quoted was from £165 to £2,000.[6]

In order to say what singers could earn from working with the smaller opera groups in Britain, we have to look at either the fee per performance or at the weekly rate of pay, and take account of rehearsal periods. Rehearsal periods could be very long if the group had an amateur chorus, so the questionnaire asked for information on how many weeks the professional singers, who would mostly be principals, were expected to rehearse. Some had *ad hoc* rehearsals, (one group reported six to nine hours a week for fifteen weeks) and others two to four weeks. Fees per performance varied considerably, ranging from £500 to £50; the average fee paid in 1988/9 was £192 per performance. However, as singers often had to give up time for rehearsals and pay their own travel expenses for quite a large number of rehearsals, their net earnings from this type of employment were comparatively low. Two of these opera groups paid singers on a weekly rate for both rehearsals and the tour of performances, as well as making travel and subsistence payments; these were similar

---

[5] In opera covers are used if a singer is unable to perform; mostly, this is due to illness or vocal problems. In British opera companies it is not normal practice for the cover to be substituted except in an emergency. A singer who does not sing is not paid for the performance; the cover is paid one fee to stand by for each performance and another if he or she goes on. This is rather different from musicals where principal singers are usually allowed a night or afternoon off and the cover would expect to go on more frequently than in opera.
[6] But see n. 4, above.

to the Equity minimum rates. Between them, the smaller opera companies hired about forty singers for around 120 performances in 1988/9. In one, singers could earn an average of £390 a week for twenty weeks in the year and in another £260 a week for forty weeks in 1988, figures comparable to what singers would have earned with Opera 80 or Glyndebourne Touring Opera.[7]

Oratorio

The source of information on fees for oratorio was from the returns made to Regional Arts Associations by the amateur choral societies in England and in the NFMS in Scotland. The data cover 3,500 individual performances by paid professional singers hired as soloists. The areas studied were Greater London, the South West, Yorkshire, and Scotland; they were chosen as being representative of the country as a whole. Data for Scotland ran from 1978/9, for Greater London and Yorkshire from 1985/6 and from 1987/8 for the South West.[8]

The societies ranged from the large metropolitan ones which were major promoters of choral concerts to small village societies putting on their annual *Messiah*; together they promoted thousands of concerts and were an important source of income for singers at all levels of the profession.[9] It should also be said at this point that to

---

[7] Two books which came out just as this book was being prepared for publication would have made my research very much easier: Cowley (1991) and Devlin (1992). They list a number of opera companies that were not included in this survey, either because they did not exist at the time or because I was unable to make contact with them. In this category the most serious omission was the City of Birmingham Touring Opera.

[8] The data analysed here were taken from application forms submitted in advance of the concert season and therefore were the intended, rather than the actual, fees. As we looked at a run of several years in most cases, it was possible to do a spot check on whether intentions had been carried out and they nearly always had. Furthermore, those societies hiring 'big name' singers would have had to book them well in advance and so the fee would be agreed in advance; this would also be true of favourite and regular singers, whom the societies also had to book in advance. In general, I would say that about two-thirds of all the fees I looked at were for specifically named singers and it seems reasonable to assume that where it was stated that Singer A was to be paid £x that the intention was in fact carried out. It may be that bias is introduced in the data by societies overestimating their ability to pay high fees in advance of the season, but this would be balanced out by other societies underestimating what they need to pay. All in all, I am confident that the data are reliable and they provide an extremely useful source of information on the range of fees for soloists in oratorio for a run of several years.

[9] Hutchison and Feist (1991) estimated that amateur choirs spent over £0.5 million on soloists in 1989/90.

qualify for a Regional Arts Association grant, the societies were supposed to pay at least the minimum fee recommended by the ISM, though the data show that this was not always adhered to.

Fee payments were ranked in £50 bands ranging from 'under £50 to 'over £850'. The results are given in Appendix 1, Table 15, which shows that in 1988/9, the median fee paid to soloists by amateur choral societies was between £150 and £199. It also shows that taking all regions together, the median or typical fee, which was between £100 and £149 in 1985/6 did not increase (or, to be more exact, did not rise out of that fee band) for the first three years under observation and then it rose in 1988/9. What is striking is how consistent fees were throughout Britain and over the period studied. In the 1987/8 season, the typical payment to soloists in each region was the same, strongly suggesting that market forces were at work equalizing fees. For Scotland only it was possible to analyse the trend in fees for the previous seven years from 1978/9 to 1984/5. The figures are shown in Appendix 1, Table 16. In 1978/9 the median fee payment was in the band £50–£99; in 1980/1, it rose to £100–£149, and stayed there for seven years.

It seems reasonable to assume in the light of the data for all four regions in Table 15 that fees in Scotland were very similar to those for the rest of Britain. Using the trend data for Scotland in Table 16 to generalize for the country as a whole for the earlier period, what this indicates is that there was no increase in the monetary value of performance fees over an eight year period from 1980 to 1988. As the Retail Price Index rose by 60 per cent during this time, the real value of soloists' fees in oratorio fell by more than half during the 1980s.

But while the median fee payment to soloists was more or less static, changes did take place at the extremes. In 1985/6 there were forty-three individual fee payments of under £50; by 1988/9 there were twenty-one. Meanwhile, in Greater London, the number of fee payments in the band over £1,000 had risen from two in 1985/6 to 23 in 1988/9. In 1985/6 the lowest individual fee paid was £20 and the highest £1,000; by 1988/9 the lowest was £35 and the highest £2,250. So, fees did rise, but not enough of them rose so as to shift up the median fee. While a few singers were doing very well, most were not. In the Greater London area, 25 per cent of the observed fees were under £100 for 1988/9, a figure perilously close to the ISM's recommended minimum of £90 for that season. The

ISM recommended minimum fee, exclusive of expenses, is intended as a guide to young musicians entering the profession. This had risen in steps from £45 for the 1980/1 season to £90 for 1988/9. My figures, however, are the typical fee paid to practising professionals, most of whom would be experienced singers not beginners.

What these data also showed was that there were two distinct ends to the market for singers. The majority belonged to the lower end of the fee scale and it was their fees which dominated the picture, increasing only slowly. Top-rank singers' fees, however, showed considerable increases: for example, in 1980 top tenors were being paid £250 per performance and by 1989 that had risen to between £1,000 and £1,500. By 1989 fees for top British singers were typically of that order, and international stars were earning much more. However, interesting though it is to look at individual singers' fees, it is very misleading to generalize from the few observations and only a few of the choral societies hired international stars.

Orchestral Concerts

The information on orchestra's payments to solo singers came from the postal questionnaire to orchestras mentioned in Chapter 3. Again, there was a considerable range of fees paid, varying from £175 to £8,000 with the average being £810 in 1988/9. Orchestras had been requested to list their highest and lowest individual fee payment and this allowed me to analyse them separately: the typical (median) highest fee was £2,000 per performance and the typical (median) lowest fee was £300 in 1988/9.

This completes the summary of information about soloists' performance fees; that was for fees paid in Britain. Fees in Europe and the USA were normally considerably higher; depending on the country, fees were one and a half, two, or even three times higher than in Britain and this encouraged many singers, both chorus singers and soloists to work abroad.

However, knowing what fees are still does not provide any answers about singers' earnings; now it is necessary to look at the other side of the equation and turn to the question of how many performances singers did. In some respects this is like asking how long is a piece of string?, since there will be variation from year to year for any one singer and variation as between singers for all sorts of reasons, personal and professional. One singer I interviewed, who usually did sixty to eighty performances a year had only done forty-five performances in the current year because of a spate of cancellations.[10] Biographies of great singers show considerable variation in the number of live performances they apparently preferred to do over the years; for example, Joan Sutherland averaged about fifty a year in mid-career (see Major 1987) while Placido Domingo averaged eighty (see Domingo 1983). The best source of information on numbers of performances was from singers themselves and their agents. Enquiries suggested that eighty to one hundred live performances a year were regarded as the maximum that could be fitted in for singers doing a mixture of opera, concerts, and recitals, with recordings slotted in between. Singers concentrating solely on concerts and oratorio might be able to do more, because they do not have the long rehearsal periods in opera. Incidentally, the ISM minimum fee was predicated on the idea of three performances a week, but it seems that few soloist singers would average 150 performances throughout the year.

The Earnings of Successful Singers

In this section the findings of my own enquiries from singers and agents about earnings, numbers of performances and fees are reported. I call these successful singers for two reasons, (1) they either had agents or were working regularly enough for me to have heard of them as singers and (2) all of them were earning above average fees for the stage of their careers at which I interviewed them. This sounds somewhat convoluted; the main point to bear in mind is that it is virtually impossible to get information about singers who have disappeared without trace after their studies or who work very little; having an agent is already a sign of relative success. The information in this section is based on the experiences of about seventy singers of different ages and career paths. I interviewed some

---

[10] If the promoter cancels a performance the singer is entitled to his or her full fee but singers often do not claim it in full, partly for goodwill, partly because they do not have to do the work and also they may fill in the time with other work. Opera singers may be in the opposite situation where they have done weeks of rehearsals (usually without payment) and then are unable to perform because of illness, etc.; in such a situation the singer does not get paid and the opera-house has to pay a cover or substitute. Singers can insure against such loss of earnings but it is very expensive and only worth while for the highest paid singers.

myself, and agents provided anonymous information about others. Agents were asked to choose several British singers whom they regarded as typical of their list but asked not to include the stars. As the information was entirely anonymous I had to rely on the judgement of the agents in question that these were 'averagely successful' singers. What is particularly useful about this information is that it establishes a link between earnings and fees.

Some of this information consists of profiles of earnings over a few years for singers at different ages and stages, showing the percentage of income that was earned in Britain and abroad, from live or recorded performance and in what branch of singing, i.e. opera or concerts. The earnings reported are obviously those from work handled by the singers' agents; if singers had other work, the earnings from that would not be included. Although the information was collected by voice type, there was no apparent difference between earnings of sopranos, mezzos, tenors, or basses: in opera it is the role, not the voice type, that attracts fee differentials—a Carmen earns an equivalent fee to a Madam Butterfly, an Otello, or a Rigoletto because they are all title roles;[11] in oratorio, usually a quartet of voices is called for (though some popular works emphasize one rather than another voice type, e.g. Elgar's *Dream of Gerontius*). The (limited) time-series information is presented in Appendix 1, Table 17 in which approximate age ranges are given; Singer 1 was the youngest at 25 and they went up in age with Singer 15 being in the mid-forties. Comparing Singer 1, a rising young star whose earnings rose from £4,000 in 1986/7 to £18,000 in 1988/9 with Singer 2, whose earnings rose from £4,000 in 1986/7 to £9,000 in 1988/9, shows the difference in the speed at which earnings can rise. By the time singers get to their mid-thirties, considerable differences between singers' earnings become apparent; Singer 15, described as a 'developing international star', experienced a rise in earnings from £60,000 in 1983/4 to £167,000 in 1987/8. It is also apparent that all the singers' annual incomes fluctuated somewhat from year to year. The effect of higher foreign fees manifested itself in earnings with those singers who did the most work abroad

---

[11] It is often thought that tenors get paid more than sopranos, sopranos more than mezzos, and so on; however, agents and opera-houses denied that this was the case with respect to leading roles. But it is undoubtedly the case that the most frequently performed operas tend to favour sopranos and tenors in leading roles and this affects the balance of earnings between different voice types.

earning the highest incomes; by contrast, singers with fairly steady UK careers in opera and with fees typical of provincial opera companies earned more modest incomes.

These figures mostly span a period of four or five years over which wage levels in the economy as a whole rose. The Average Earnings Index (Department of Employment 1990) measures the average annual increase in earnings;[12] between 1985 and 1989 this was 40 per cent. The Average Earnings Index is given in Table 17 and it can be used to make rough calculations of the extent to which these singers' earnings kept pace with earnings in other occupations. For example, between 1985 and 1986 average earnings in the economy as a whole rose by eight per cent; therefore a singer earning £20,000 in 1985 would have to earn £21,600 (an 8 per cent rise) by 1986 to keep pace with earnings elsewhere in the economy. (So Singer 7's income actually fell relatively speaking between 1984/5 and 1985/6, since it stayed at the same monetary value.)

The information about the earnings of successful singers is valuable in its own right but we know nothing about how representative these singers are of the profession as a whole. What they in fact represent is typical averagely successful singers, but not, it should be said the most successful singers. Generally speaking, informed opinion has it that good singers would be expected (in 1990) to earn between £50,000 and £100,000 a year. The top British stars who are established international singers seemed to earn even more, grossing at least £200,000 a year.

Cross-Section Information on Singers' Earnings

A different type of data about singers' earnings is of a cross-section of the singing profession for one particular year. In Appendix 1, Table 18 data on twenty singers' earnings, both gross and net, are presented for 1988, together with their usual fee and the number of performances they did that year. In addition, a rough estimate is given of what percentage of their work was undertaken in the UK. The singers covered the whole range of the profession from top to

---

[12] The base of the Average Earnings Index was changed in 1985 and it is a rather complicated matter to go back to 1984. The season over which singers usually work is Sept. to July, so that the relevant annual average figure falls between these dates. The only easily available figure for the Average Earnings Index in 1989 was the July one.

bottom—singers who performed part time and taught part time, did concert and sessions work, recitals, oratorio, opera, musicals, and cathedral work, and covered every age and career stage. The information in Table 18 demonstrates the wide variation in earnings, fees, and numbers of performances that occur in the singing profession. The responding singers earned between £100,000 and £1,000 in 1988; fees ranged from £20 to £900 and the number of performances from four to 452. This shows how difficult it is to generalize about singers' earnings. In some cases fees were irrelevant as the singer was employed on a salary in an opera chorus, cathedral, or musical. Singers with very high numbers of performances were performing over eight times a week, probably in a musical or a cathedral choir. In other cases where singers performed very little it was because teaching was their main source of income (or even private money in a couple of cases). Low fees suggest that the singer was doing concert and sessions and/or church work, or was charging a low fee for solo work (the ISM minimum solo performance fee in 1988 was £80 up to July and £90 after). Median earnings were £12,000 which is consistent with the ISM's survey of musicians discussed earlier on.

The Earnings of Singing-teachers

Information on the earnings of singing-teachers comes from the postal questionnaire to singing-teachers conducted in 1989, details of which were given in Chapter 1. In order to get some idea as to the financial side of a career as a singing-teacher, teachers were asked what they charged for lessons and secondly, what their tax-deductible expenses were for private teaching.

In 1989 the average charge per hour for a singing lesson was £14, which was the mean of a range from £45 down to £8. It is interesting to compare this with the results of my questionnaire study of all institutions of higher education in which singing was taught. In the music colleges, singing-teachers were paid between £12 and £14 an hour; in universities the average was £12, and £14 in polytechnics and Colleges of Higher and Further Education. Thus the average fee of £14 for private teaching was very much in conformity with those rates. However, teaching at home necessitates a certain amount of extra expenditure and that is why the questionnaire asked for information about tax-deductible expenses; 40 per cent of respondents answered this question; their average expenses were £1,000.

What did singing-teachers earn? Working an average of twenty hours a week at a rate of £14 per hour, they earned on average £280 a week. If they were to work fifty weeks a year they would have earned a gross income of £14,000 a year from teaching singing in 1989; however, there are various reasons why it is unlikely that that was possible. Those who were teaching in institutions of higher or further education would, unless they were some of the very few with a permanent contract, be paid for around thirty-five weeks of the year. Although private teaching can be carried on all through the year, pupils cancel lessons, have colds, etc. and so even if the teacher is expecting to teach a certain number of hours, he or she may not get paid for every timetabled hour. Furthermore, many teachers do not charge all pupils their full rate, so the figure of £14 overestimates the average payment. Nevertheless, taking account of these factors and the costs of running a private teaching practice, it seems likely that the average singing-teacher could earn about £10,000 a year in 1989 for an activity that they pursued for half their working time. For some this was a useful part-time or retirement income. For others it supplemented a performing career. Looking at fees paid to singers by choral societies and the like, this showed that in 1989 the typical fee was between £150 and £200. Thus to equal the earnings from teaching singing half time, a singer-cum-teacher had to undertake 50 or so concerts a year. So by teaching twenty hours a week and performing once a week, the average singer/singing-teacher of our survey would have earned £20,000.

Gross and Net Earnings and Singers' Expenses

The cross-section data on singers' earnings in Table 18 showed a considerable difference between the singers' gross earnings and their pre-tax net earnings. The difference is accounted for by the singers' expenses. Expressed as a percentage of gross income, expenses ranged from 83 per cent to 10 per cent, some of the highest ratio of expenses to earnings being for the lowest paid singers. One singer provided detailed information on his/her tax-deductible expenses and gross income for twenty years from the beginning of his/her career at the age of 23. The figures are presented in Appendix 1, Table 19.

This singer was born in the early 1950s and has had a fairly steady career. He/she has regularly worked with most of the British opera companies and at one point in her/his career did a great deal of foreign work, but preferred for personal reasons to spend more time in Britain; earnings and tax-deductible expenses are given separately. The information runs from 1975/6 until 1987/8; the relevant run of the Retail Price Index is given so that the reader can calculate whether real income had risen or not. The singer's income rose from £2,000 to £15,000 during this period, i.e. it rose seven and a half times.

Over these thirteen years, the cost of living, as measured by the Retail Price Index, trebled, so by the end of the period this singer's real income had risen. But his/her expenses had risen ten times. Deducting tax-deductible expenses from gross income to obtain net income, the picture looks very different; even in monetary terms, his/her net income hardly rose and certainly did not keep pace with rises in the Retail Price Index.

How does this singer compare with other singers for whom there is information over a period of time, i.e. those included in Table 17? At age 25 he/she earned £2,000, which was the equivalent of £4,200 in 1986/7, so this singer began with much the same earnings as Singers 1 and 2 in Table 17, though his/her earnings rose more slowly. He/she was not so successful financially and hence is more representative of the profession as a whole than those 'successful' singers. But how representative of the whole profession is this singer? She/he was being paid average or typical fees in opera in Britain and was doing around forty performances a year, which was probably standard for the stage of his/her career and age. He/she was earning £12,100 at the age of 34, i.e. the same as the median in the ISM study, but that study represented older and more successful musicians. So the singer in Table 19 was probably more successful than many others, and the fact that he/she had worked with nearly all the British opera companies and was successful abroad at one stage reinforces that judgement; she/he did solo work, while the mass of the profession probably do chorus work of one kind or another. On the other hand, singers in opera choruses, at least, were paid £10,800–£12,800, i.e. in the same range as this individual. All told, then, this singer was probably pretty representative of the singing profession.

However, the figure on tax-deductible expenses provided by the singer in Table 19 included such items as mortgage repayments and insurance, i.e. all expenses were not just those connected with singing. To define what expenses associated with singing actually are, and what the tax position is with regard to them, is quite complex.

Singers' Expenses and Taxation

Gross income includes payments made to singers for travel and subsistence as well as the payment for their services, i.e. the fee or salary. Tax-deductible expenses of singers, however, include many items besides travel, hotels, and other such subsistence expenses. Exactly which are expenses tax-deductible depends on whether the singer is paying Schedule D or Schedule E (PAYE) taxation, i.e. whether self-employed or employed; for example, if you are employed your employer is deemed to provide you with a place of work and you cannot count your music room as a tax-deductible expense, though you could do so if you were self-employed. Furthermore, singers with a turnover of £23,500 (in 1990) had to register for VAT. This is not the place for a detailed discussion of taxation and only the most general aspects are mentioned here as they relate to singers' earnings.[13]

Expenses that are peculiar to singers and which continue almost throughout their careers are the time and direct costs of learning solo repertoire, the direct costs being the costs of singing lessons and coaching with a repetiteur. Singers may learn roles, songs, concert pieces, etc. on their own and/or with a coach; many work with a coach, because unlike instrumentalists, they often have to learn hundreds of lines of words by heart in addition to the music and find it a better discipline to do so with a coach. In addition they have to maintain their voices by regular vocal practice, singing lessons, and keeping in good general bodily health. The time cost of all this is a very important item, because a great deal of time is involved and because time has an opportunity cost; learning time for freelance singers is not usually paid for by the employer in Britain and the singer is expected to be word- and note-perfect by the first rehearsal. One very experienced recitalist reported that it took a whole morning to learn a long and not very well-known

---

[13] Equity is able to advise members about Inland Revenue's decision in 1990 to bring singers working in theatres under PAYE rather than Schedule D and there is a useful introductory ch. on Income Tax and National Insurance, as well as on VAT, in Ford (1986), a book which is very worth while for its other chs. as well.

Schubert song and then she/he looked at it again for half an hour a day for four or five days before the concert; there could be twenty or so songs in a recital (though not all of them would be new to the singer). Opera singers report that it can take several months, even a year, to really learn a major operatic role. Another special area is Early music, where singers often have to struggle with photocopied original manuscripts and medieval French or Latin; this is also very time-consuming. Singers on regular contracts do not have to bear the costs of learning themselves and are usually paid for learning time. Rehearsal time is another time cost to most principal singers. With some opera companies in the 1980s having rehearsal periods of six to eight or even twelve weeks, this imposes a high opportunity cost on singers. However, opportunity costs cannot be charged against taxes, though direct costs may be.

Singers were asked about their expenses for coaching and singing lessons and figures of £1,000 to £2,000 in 1988/9 were typical for singers at all levels. These may increase somewhat as the singer's career advances, because they get bigger roles to learn and probably work with more expensive teachers and coaches. On the other hand, they can exploit their accumulated repertoire and thus learning costs may fall after a time. Young singers have to bear high learning costs at the beginning of their career—one reason why being a cover is useful, because they get help with learning roles. (Young singers are often offered roles that are not in the usual repertoire, probably because older singers have no financial motive to learn something they will never be able to use again. This imposes even higher learning costs that cannot be spread over a large number of performances.)

Other expenses are evening clothes, again a big item for young singers. A woman's evening dress of the type expected of an established performer could cost £500 to £1,000 in 1990 and it was not difficult to spend more. One singer had spent £12,000 in a year on clothes, shoes, and jewellery for performance purposes only! Another big item for beginners is having a set of professional photographs made which can cost several hundred pounds, and many singers in addition have professional tapes made, which is also very expensive. However, these are all tax-deductible expenses. Singers can also count as non-taxable expenses items ranging from Ear, Nose, and Throat specialists' fees, agents' fees, cosmetic dental work, and contact lenses to getting their hair cut and purchasing the *Radio Times*!

Singers who are self-employed (Schedule D) undoubtedly benefit from being able to treat many expenses as tax-deductible that would not be possible for singers paying PAYE (Schedule E), especially if they have a good accountant, but, on the other hand, self-employed people have to pay all their own National Insurance and pension contributions. However they are taxed, singers have genuinely high expenses, some of which are not tax-deductible; chief amongst these is the time cost of learning and rehearsing, as already mentioned, but there are other items as well; for instance, women singers with children could not (in 1990) deduct the cost of a live-in nanny against tax even though they could not work full time without one. Furthermore, singers working abroad, often in many different countries, have to deal with the diverse taxation rates and practices that pervade even the EEC, claiming back withheld taxes which are automatically deducted in some countries and never seeing again the statutory pension contributions that are required in a few. What look like glamorous foreign fees can often dwindle when the realization dawns that not only are tax rates, etc. higher abroad but so are prices. The result is that, in general, singers say they typically come back home with about half of the fee they are paid for work abroad. For work in Britain, singers reckoned that 35–40 per cent of their gross incomes went out in expenses. Thus pre-tax net incomes were about 60 per cent of gross income.

What this means is that a freelance self-employed singer earning a gross income of £20,000 has a net pre-tax income of £12,000. So, a singer in an opera chorus earning £12,000 a year, with the employer's contribution to National Insurance paid, being paid for rehearsal and learning time, and having a place of work—all the items that freelance singers have to pay for directly or indirectly themselves—is as well off in net terms as a self-employed singer earning £20,000. In order to earn £20,000 a year, a singer would have to do forty performances a year at a performance fee of £500 or 100 at £200—perhaps not untypical of many British singers. A performance fee of £500 for a principal role was reported as being the norm in British opera companies in 1989 and in oratorio £199 was the top of the median fee range. Gross earnings of £20,000 were also estimated to be the typical singing-teacher's income in 1989. Therefore, by taking account of singers' expenses, a consistent pattern of earnings in different branches of the singing profession emerges.

The fact that singers have high tax-deductible expenses does not mean that they do not have to pay taxes—they pay taxes like every other worker. What it does mean is that they pay proportionately less tax on a given income. An example will make this clear; assume that the only tax rate is 25 per cent and that all net income is taxed. Also assume that an employed singer paying PAYE taxes has no tax-deductible expenses:

Self-employed singer paying Schedule D tax:
   Gross earnings   £20,000
   Tax deductible expenses   £ 8,000
   Net pre-tax income   £12,000
   Tax at 25%   £ 3,000
   After-tax income   £ 9,000
Employed singer paying Schedule E tax:
   Gross earnings   £12,000
   Tax-deductible expenses   £   0
   Net pre-tax income   £12,000
   Tax at 25%   £ 3,000
   After-tax income   £ 9,000

This example shows that while the employed singer took home 75 per cent of his or her gross income, the self-employed singer took home 45 per cent.

Summary of Findings on Singers' Earnings in Britain

Data on singers' earnings have been assembled from a number of different sources: surveys of Equity and ISM members, salary payments to singers on regular contracts, information from agents, and my own surveys and interviews.

Data were also obtained on fee payments to freelance singers and an attempt was made to combine this information with figures on the typical number of sessions worked or performances given, in order to roughly calculate typical incomes of singers working in different branches of the singing profession, and to compare gross and net earnings. In most cases, the information is not systematic and it cannot be claimed that it is objective. On the face of it, there seems to be enormous variation in singers' earnings and it is very difficult to generalize.

But on closer inspection a pattern does emerge. To begin with, there is a certain consistency in the figures for median earnings of salaried singers: those working in the permanent opera choruses earned between £11,000 and £13,000 (in round numbers) in 1988/9; the BBC Singers earned a basic salary of £13,500 in 1989; singers in the chorus of musicals earned a basic £13,000 in 1989. The ISM survey of its members (all types of musicians, not just singers) who were freelance soloists showed a median income of £12,000, as did those in my cross-section data; the 1987 Equity survey suggested much lower earnings, but that was for a wide range of performers and singers were shown to be among the higher earners in Equity; young principals in opera companies were paid between £9,000 and £14,000 in 1989. So we are able to arrive at a consensus figure of something between £10,000 and £14,000 for 1989. The mid-point of £12,000 seems to be not an unreasonable one to settle for.

The greatest difficulties were, predictably, with freelance singers' earnings and there generalization is likely to be misleading. Data from agents on earnings of their typical singers showed considerable differences, even for singers in the same age group. However, that information is valuable if only because it showed that some successful singers (soloists) with agents were earning as little as £9,000–£15,000 in 1989, i.e. less than the average singing-teacher. They also showed how little young singers of 25 years of age earned at the start of even a promising career. Those data, and the cross-section data, were particularly valuable because they linked earnings to fee payments, and showed that singers working abroad generally earned more than those working mostly in Britain.

How did the earnings of singers compare with those of other workers? In 1989 average non-manual earnings were £19,500 and £10,800 for men and women respectively in the age group 40–9. Thus, if we take £12,000 as the typical figure for singers in 1989, women did comparatively well in the singing profession but men earned significantly less. But for freelance singers this figure for gross earnings considerably overstated pre-tax net earnings, and so it seems that this is not a fair comparison to make, given that the majority of non-manual workers could be expected to have lower work-associated expenses. Taking that into account would increase the gap between average earnings of non-manual workers and singers even more.

On the other hand, it could be argued that singers are generally younger than other non-manual workers. The age range 40–9 was chosen because that is the right one for mid-career comparisons;

many singers are really only getting themselves established in their thirties and some voice types, particularly light sopranos, could be past their best in their fifties. The earnings of non-manual workers in all occupations vary a lot by age. Non-manual workers of 25–9 years of age earned on average in 1989 as follows: men £14,404; women 10,712 (Department of Employment 1989; see Appendix 1, Table 20). What is called the age-earnings profile usually rises up to the age of 40–9 and then tails off. For example, according to figures from the General Household Survey, in 1987 the average earnings of men with an Arts degree were £11,505 for the age range 25–9; £11,875 for the age range 30–9; £19,005 for the age range 40–9; and £12,848 for men aged 50–9. Thus average earnings were over £7,000 a year higher for male Arts graduates in their forties than for those in their thirties. However, these figures are for people working in occupations where salaries rise with age and experience; as we have seen, singers' salaries, for example, in the permanent opera choruses only rise slightly if at all, with age. Concert and sessions singers are paid the same rates whatever their age. Without systematic information on singers' earnings by age, it is not clear which age group would be chosen for purposes of comparing the earnings of singers with those of other workers. It is obvious that the conclusion of the debate about whether or not singers earn comparable incomes to other workers depends on the age group chosen. If singers were all 25–9 years old, their salaries would compare favourably with those of other workers, especially those of female singers.

A different problem is presented by inflation. If comparisons are not made strictly on the basis of the same year, inflation may interfere with the comparison. I have attempted to do this as carefully as possible, but even one year could make a difference. In the 1980s the Average Earnings Index rose considerably—40 per cent from 1985 to 1989; in only one year, 1987 to 1988, it rose 14 per cent (see Table 17). So earnings of £12,000 in 1987 would have risen to £13,680 in 1988. However, it was argued in Chapter 4 that, from the evidence of the Peacock study (Peacock, Shoesmith, and Millner 1982), payments to singers had not risen at the rate of the Average Earnings Index. Indeed, whatever the initial differentials between the earnings of singers and other workers, they must have increased during the 1970s. This is likely to have been the case in the 1980s too.

To conclude this summary: the attempt has been made to establish whether or not singers earn as much or less than other comparable workers. However, the argument is conducted, the sad fact is that the data presented here are not accurate enough to make a proper comparison. In several countries, particularly Australia and Canada, major surveys of artists of all types have been undertaken, which provide statistically reliable data on the earnings and other economic and social characteristics of specific artistic occupations. In other countries, notably the USA, data on occupational earnings are collected in the Census of Population. This has enabled economists to systematically compare the earnings of artists with those of other workers. One such study by Filer created somewhat of a stir by showing that when all the relevant variables, i.e. age, sex, race, educational attainment, were carefully standardized, artists did not earn less than other workers—the 'starving artist' was a myth (Filer 1986). Unfortunately, the option of doing this type of work in Britain is not available.

Does the Market for Singers Work?

In a labour-market that is working freely, that is, where only the forces of supply and demand determine the wage, wage rates for similar workers would equalize. 'Similar workers' mean workers of comparable age, experience, and education, the latter being a measure of ability. Extensive studies of labour-markets have shown that even in the absence of institutional factors, such as trade unions fixing wage rates, wage rates do not always equalize after account has been taken of age and educational attainment. There are two main reasons for this; discrimination and segmentation. Discrimination is either sexual or racial. Segmentation is the term used for non-competing groups in the labour-market; a segmented labour-market is one in which there are basically two types of workers those in the 'internal' labour-market in which there is a career ladder with regular possibilities for internal promotion, higher earnings and other benefits and those in the 'external' labour-market, where there are limited job prospects, high unemployment, and casual work. The significance of this theory is that the external labour-market does not compete with the internal one.

How does the market for singers fit these characteristics? Does it work freely, do rates of pay equalize, and is there discrimination and segmentation?

Equalization of Rates of Pay

The evidence presented in the previous section suggests that there is equalization of pay in the market for singers. This is more apparent at the level of chorus work, i.e. in the opera choruses, concerts, and sessions, the choruses of musicals and the like, but also for young principals and those working in the smaller opera companies. These are all areas in which Equity sets minimum rates of pay and the question therefore arises as to whether supply and demand are at work and determine payments or whether the market accepts imposed fixed rates, negotiated by collective bargaining rather than being determined by market forces alone. Besides Equity, the ISM also has minimum rates, which are accepted by the BBC and by organizations in receipt of public subsidy, such as the amateur performing societies. The evidence has shown that in the latter case, at least, these minimum rates are undercut by some singers, which suggests that market forces are at work. But, although the Equity minimum rates are strictly adhered to, that still does not ensure that work is available at those rates; in other words, singers' earnings, which depend both on the rate of payment and the amount of work offered at that rate, may still be subject to market forces even in the presence of fixed rates. What seems to me to be the situation is that there is an abundant supply of singers but insufficient demand to enable them all to be regularly employed, hence the low earnings of many singers. The market for singers works according to supply and demand, even with fixed rates of pay. That supply exceeds demand at these rates is to be seen in high rates of unemployment and underemployment (singers who only work part of the time) as witnessed by the Equity Survey, and also in the fact that only one of the employers I interviewed paid (marginally) over the relevant Equity minimum rate.

What about soloists—the successful singers? Here other factors are at work. Highly talented singers, the stars, are in short supply. They do not compete with the bulk of the singing profession. Almost by definition stars cannot easily be substituted for, and demand for their services is accordingly very high. They inhabit a world much closer to the internal labour-market, suggesting that there is some segmentation in the market for singers.

Segmentation

If there is segmentation in a labour-market this is because groups of workers do not, cannot, compete effectively. This is partially the case in the labour-market for singers, in so far as the stars do not experience real competition. The dividing line is great talent. However, there are no absolutes here. In principle at least, all trained singers could become stars, indeed they mostly believe they will do so. It is only through work experience that singers can discover their chances of success, and it is only by seeing singers perform that those who hire them can properly evaluate their ability. Competition (including singing competitions) is an inevitable part of the process of discovering talent for both singer and employer. And there is a well-developed network that collects and disseminates information about talented singers. Thus, there is potential competition between singers at all levels.

There are some areas in singing in which effective competition is limited by the fact that substitution cannot easily take place. There are groups of singers with specific attributes who do not compete across the board. Voice type is the obvious example of this; sopranos do not compete with tenors (though mezzo-sopranos may have to compete with countertenors). Singers who cannot sight-sing or who do not have perfect pitch cannot compete with those who do, and so could not easily enter the market of contemporary music. Singers with little vibrato in their voice or small voices do not usually compete with grand opera singers. So there are different sub-markets for singers based on special characteristics. But that is not to say that if there were a striking shortage of singers with one or another characteristic, no-one could fill it. And it is competition and the chance of higher earnings that is the motivating force.

The more easily one singer can be substituted for another the smaller the earnings differentials between sub-groups will be. At the limit, rates of pay will equalize. That they do so for chorus and similar work is not surprising, because substitution is easy and the supply of substitutes plentiful.

Sexual and Racial Discrimination in the Singing Profession

Was there any sign of either sexual or racial discrimination in the earnings data? We would expect discrimination to show up in earnings because, in the economy at large, women and people from ethnic minorities in general get less work and are paid at lower wage rates. Neither the Equity survey nor the ISM survey produced information about ethnic minorities' earnings; however, the Equity survey data were analysed separately to look for evidence of earnings differences between women and men, and the analysis confirmed what many women in Equity had felt to be the case for some time, that women get less work measured in weeks worked than men, and consequently earn less (reported in Equity 1990). There is no evidence that women in the singing profession are paid lower rates of pay and Equity rates are, of course, the same for men and women. In the cross-section data presented in Table 18, half the singers earning above the median were women, though over two-thirds of those earning below the median were women (60 per cent of the sample were women); however, the median of women's earnings was lower than that of mens' (£10,000 as compared to £14,000).

It is very difficult to establish why women do worse than men, because in the singing profession work is very evenly balanced as between male and female voices; choirs, if anything, tend to have a few more women than men in them, though the cathedral choirs are an exception here. Solo parts in oratorio are usually evenly balanced in a quartet and most principal roles in operas favour women. I did find some evidence of men getting more work in the form of overtime hours in opera (in one major opera company the men in the chorus did 135 performances in 1988/9 compared to 115 for women, because the operas in that season's repertory called for more input from the male chorus). In general, though, the demand side does not heavily favour men. The problem would seem to be more on the supply side, with many more women than men training as singers both in institutions of higher education and privately. In Chapter 1, it was estimated that two-thirds of all singers training were women. Whether or not there is an oversupply of female singers is as difficult to establish as the question of a general oversupply of singers. It is widely held that there are far too many sopranos. An excess supply of female singers would certainly cause the general level of their earnings to fall. Another factor that undoubtedly comes into play in the singing profession is that women are less prone to leave it than men, partly because their alternative work opportunities do not pay as well (therefore their opportunity cost is lower) and partly because it suits some women to work part time—for example, if they have children—and they may not feel they have the same economic pressure as men, a widely held sexist view, which is widely held by women and men alike! This viewpoint may in fact underlie the apparent sexual discrimination in the awarding of bursaries and scholarships, of which I found some evidence; although considerably more women than men apply for awards and enter competitions, the success rate is more or less even as between the sexes, i.e. proportionately fewer women applicants or entrants succeed. It could be that young men are regarded as being more needy than young women. It could also be that women are more unrealistic about their chances of success than men, and, indeed, that is why they go in for a risky profession like singing in the first place.

These are very tricky issues and by no means easy to analyse with the evidence available, if at all. This is perhaps even more the case with racial discrimination. In a profession in which a great deal depends upon luck and upon the taste and preferences of those who audition singers, and where there are few objective criteria for excellence, it is even more difficult than usual to establish whether or not there is racial discrimination. On the one hand there are a number of highly successful ethnic-minority singers who sing the whole range of the principal operatic oratorio and concert repertory. On the other hand, there is strong hearsay evidence of a few opera producers, organizers of choral societies, and other types of promoters rejecting ethnic-minority singers, possibly because of their ethnicity. There are ethnic-minority singers in several British opera companies and in the 1980s, two choruses, those for the opera *Porgy and Bess* at Glyndebourne and in the musical *King* were both formed of exclusively Afro-Caribbean singers with *Miss Saigon* employing mainly Asians. What is needed in order to establish whether or not ethnic-minority singers are being discriminated against is a set of data on earnings that are representative of the profession as a whole and that distinguish a number of features of social and educational background, including ethnic origin.

Enough has been said to confirm that the market for singers, by and large, works through competition to equalize net earnings for comparable singers. But we are still left with differential earnings, which cause the distribution of earnings in the singing profession to be uneven. The cross-section data on singers' earnings and earnings of successful singers in Tables 17 and 18 showed that there is enormous variation in the earnings of singers. This is a topic of considerable interest to economists, and several theories have recently been advanced to explain the uneven distribution of earnings in certain professions.

The Distribution of Earnings in the Singing Profession

The earnings data in Table 17 show that the distribution ranged from gross income of £1,000 at the bottom to £100,000 at the top. If incomes are classified into bands of £1,000–£2,000, £2,000–£3,000, and so on, it is possible to draw up the distribution of income relating the frequency with which these income bands occur in the data. If incomes are normally distributed the income distribution will appear as in Fig. 5.1 and the average income will fall in the centre of the distribution. However, it has long been realized by economists that in many professions, particularly those in which most workers are self-employed, instead of being normal, the distribution of income will be skewed as in Fig. 5.2, showing that a large proportion of people in the profession earn relatively low incomes while a few earn very high incomes. Clearly, the distribution of earnings in the singing profession is skewed and it is for this reason that median rather than average income figures were used in the earlier discussion.

Various explanations have been put forward by economists to account for this type of income distribution. Starting from the basic point of view that what people get paid for is the value their contribution makes to the output of goods and services, the question is why does one person performing a particular service, e.g. singing

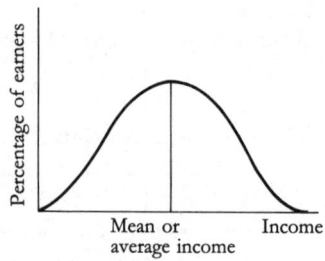

Fig. 5.1. Normal Distribution of Income

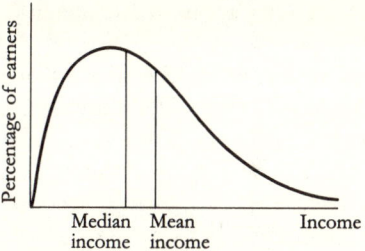

Fig. 5.2. Skewed Distribution of Income

a specific operatic role, get paid a very different fee than another. When Domingo sings Otello at Covent Garden he gets paid perhaps ten or fifteen times as much as a singer doing the same role at the Welsh National Opera or the English National Opera. Indeed, the same singer may on occasion sing the same part at very different fees, say in Britain compared to France or Italy. Why? One set of explanations has to do with the pricing of the final service itself, i.e. with the ticket price charged for the performance and the contribution that an individual performer may be deemed to make to that; differential levels of arts subsidies may also play a part in this; these issues were discussed in Chapter 4. But there are other explanations as well.

All singers are not equally talented and there is a relative scarcity of talent as one, as it were, goes up the scale. But the connection of talent to earnings is not so obvious. For if one singer earns ten times more than another, are we to conclude that he or she is ten times more talented? Even without an objective measure of talent we would surely say no. Therefore it is not relative scarcity of talent alone that accounts for earnings differentials. One economist, Sherwin Rosen, has attributed the earnings of superstars to the development of media technology which has allowed singers, entertainers, athletes, and the like to reach ever larger markets without significantly increasing the costs of production; the effort and costs of doing a concert are the same whether ten or ten thousand people are in the audience or one or one million people buy a record, and so the return to talent is greatly multiplied by the market size the performer can reach (Rosen 1981); this means that small differences in talent lead to multiplied differences in earnings and to the tendency for a few big names to dominate the market. This causes the distribution of incomes across the profession in question to be skewed.

These ideas have been developed by two other economists, Adler (1985) and MacDonald (1988); Adler shows that these predictions hold even if we cannot measure differences in talent between performers (or indeed, if there are no differences in talent), because consumers, whose leisure time is valuable, make choices about which performers they favour by choosing stars whose talent is widely acclaimed, rather than invest a lot of time in finding out about less well-known performers. I would add that this is what promoters do too, in order to minimize their search costs. Again, this behaviour leads to the market being concentrated on a few top performers whose high incomes dominate the profession. Another type of explanation is offered by MacDonald, which suggests that the cause of the uneven distribution of income originates on the supply side rather than on the demand side; the bottom end of the income distribution is dominated by young beginners, who work for very little (and less than their opportunity cost) while they acquire performing experience, after which they either drop out as they get bad reviews or are unsuccessful at getting reasonable work or continue in the profession if well reviewed and are reasonably successful. Older and more experienced professionals who survive the course attract bigger audiences and command higher fees, but are fewer in number than younger lower paid performers, hence the skewed income distribution. This latter explanation seems appealing and has a ring of truth about it as far as the singing profession is concerned, but it does not really explain the type of earnings differentials that we found in Table 18, in which there are singers of the same age earning very different incomes; nor is it true that only younger singers are badly paid—some older ones are too, presumably the less talented.

My own explanation is different again. What I believe to be the case is that two things work in tandem for the top stars; the performance fee and the amount of work that a singer is offered increase together; indeed, the higher the superstar singer's fee the more work she or he is offered, because the fee is used as an index of talent and serves as a source of costless information. The more a singer is in demand, the more promoters will demand his or her services. Up to a point, this will result in the singer doing more work, but when the absolute maximum number of performances that a singer can do is reached, the fee will go on rising as demand increases relative to a fixed supply. As both the fee and the number of performances rise together, the multiplicative effect on earnings results, as in Rosen's explanation, in very much higher earnings at the top of the profession and conversely, low earnings, low performance fees, and few performances a year at the bottom end. But the technological constraints of live performance (and there are as yet no superstar singers who do not perform live, however many records they make) open up a mid-field in the profession of people who are not on their way up as stars but who have good, solid careers (see Towse (1992*a*) for a more detailed treatment of this topic). The advantage of this explanation is that it avoids the convoluted thinking about talent and quality in Rosen's approach and the supposition made by MacDonald that performers drop out when they get bad reviews. One only has to read a few singers' biographies to realize that if they believed their first reviews or audition panels there would be few singers left in the profession! However, what remains to be explained is how superstars are created. Whatever the explanation, and they are not necessarily mutually exclusive, it is clear that the distribution of earnings in the singing profession is skewed, showing that many singers earn low incomes, while a few have exceedingly high incomes. The singing profession has this in common with other professions in which information about the quality of people's work, i.e. talent or outstanding ability, is difficult and costly to obtain.

Conclusion

In this chapter, information on singers' earnings has been presented and the question raised as to whether or not earnings are determined by the interaction of supply and demand. It was argued that the market for singers in Britain works through competition despite fixed minimum fees and wage rates. But because singers are not equally talented, competition between the less and the more talented is not completely effective. Along with other labour-markets in which one worker cannot easily be substituted for another, the distribution of earnings is skewed, with the stars or superstars earning vastly more than the bulk of the profession. In particular, the market in such situations has a tendency to concentrate on a few high earners, partly because of economies of scale due to developments in media technology and partly because the cost of information about talent encourages the use of high fees as an index of quality. High-earning singers command high fees and a lot of work and so differences in talent are multiplied out of all proportion.

The possibility of high earnings is an incentive to enter the singing profession, even if the probability of becoming a star is low. But those who do not succeed, and that is clearly the majority, earn relatively little, and less age for age than those in other occupations with comparable educational requirements. How earnings influence the decision to undertake the training that is needed to become a singer is the next topic to be discussed.

6
The Rate of Return to Training as a Singer

In this chapter we draw together information on the costs of training to be a singer and the financial returns in the form of earnings from employment in the singing profession which have been presented in previous chapters. The conclusion of the last chapter was that singers' earnings are comparable to average non-manual earnings for the age group 25–9 but are lower than those of older workers. In Chapter 2, it was shown, however, that the true cost of a singer's training is higher than the cost of training other graduates. This was partly because of the direct costs of formal training but also because of higher indirect costs, particularly those of forgone earnings. The implication of these findings is now spelt out in more detail.

Training is an investment in human capital and the task here is to calculate the rate of return to that investment, and to discuss whether or not it is a good financial proposition to become a classically trained singer. How one uses such information is a separate matter; it might be used by individuals in thinking about careers and by governments in deciding how much subsidy should be given to institutions that train singers. Studies in the economics of education have shown that, in general, students do respond to financial incentives in making their choice of which subject to study in higher education and what career to apply it to subsequently. A study of nearly 3,000 school children done in 1977 showed that their perceptions of earnings were pretty accurate (Williams and Gordon 1981). Singers may be different from other students and financial considerations may not be uppermost in someone's mind when they decide to become a singer (though they certainly might be in the parents' minds!). In making these calculations it is not being suggested that this is how young people should behave; there is, in fact, no 'should' about economics at all. The aim is rather to provide a slant to other aspects of the debate about training singers. To date, this debate has not been well informed about the economic aspects of the training and employment of singers.

Training as Investment in Human Capital

Since the 1960s economists working in the fields of health and education have viewed human beings as investing in themselves through the formation of human capital. The time and resources taken to build up *mens sana in corpore sano* constitute the amount of the investment, and everyone has some element of human capital embodied in them. Those who have taken out more time to train, study or keep fit, and who have received or spent more resources in the form of a place in higher education or private singing lessons, embody a higher investment in human capital; the longer the education a person has received, the greater the investment in human capital. Capital in economics is deferred consumption; in a Robinson Crusoe economy (much beloved of economics teachers!) when Robinson Crusoe decides to set aside the day to make a fishing-net rather than catch fish with his bare hands, he gives up that day's fish dinner (consumption) for the option of catching more with his net (the capital) thereafter.

Similarly, students forgo earnings and present luxuries for the years during which they undergo higher education because they hope to earn more afterwards than they otherwise would do. (However, because it is not all torture, and students may actually enjoy higher education, not all of it can be regarded as investment; some element will be consumption.) The pay-off to the investment in human capital is partly pecuniary—higher pay—and partly non-pecuniary—the chance to do a job you enjoy, meet interesting people, enjoy social status, respect from the community, and so on; this latter group of rewards are sometimes called 'psychic income' by economists and may be an important consideration in the choice of career.

In general, the way our society and economy work is to reward people more who have made a higher investment in human capital, among other ways by paying them higher incomes; thus it is usually the case that the more years of formal education people have, the higher are their incomes. The relationship between earnings and the length of formal study have been queried in the arts professions

(Filer 1986) but it seems reasonable to suppose that there would be a positive relationship in singing.[1] The issue of non-pecuniary benefits, psychic income, is more difficult to deal with, mainly because it is extremely hard to know how to measure such rewards but also because people tend to treat any gap between the pecuniary earnings of artists and the earnings of comparable professionals as psychic income (Withers 1985). It is all too easy to say that artists are dedicated to their work and do not expect to earn as much as others (and, that anyway, they do not just work for money and enjoy what they are doing too much to worry about how much they are paid). It is precisely because of this that it is important to discover whether the myth of the starving artist is or is not a myth. If artists do earn less than other professionals in pecuniary terms, there is a tendency, both on the part of economists and non-economists, to rationalize their lower earnings by saying they are equally well off by virtue of job satisfaction and the other psychic benefits. This argument is even used to justify the lower earnings of artists. Pecuniary earnings normally increase with the number of years of formal education and, as it is only possible to measure pecuniary earnings, the true benefits to the individual of investment in human capital are underestimated by the (unmeasurable) amount of psychic benefits. The return to the initial investment is the stream of annual earnings over the person's whole working life during which it is assumed that the benefits of education last.

The rate of return is the percentage amount by which discounted total lifetime earnings exceed the discounted value of the outlay on the investment, that is, the total cost of training or education. In principle, this is exactly the same as the rate of interest on a deposit with a bank or building society or on a bond or other means of

---

[1] In Ch. 5, Filer's study, which showed that the starving artist is a myth was briefly mentioned. In this study, he also found a negative relationship between the length of study and earnings in the arts. His explanation for this is that artists are more likely to benefit from learning by doing than by formal study in an academic environment; moreover, it may be that it is less gifted artists who go on studying longer hoping to gain by nurture what they are denied by nature. Throsby (1986) found a weak, though not negative, relationship between years of education and earnings in his study of artists in Australia. I have found the same thing in my study, of artists in Wales (Towse 1992b). These results are for all arts professions taken together and I personally doubt that this would be the case in the singing profession. Indeed, it is often only the most gifted singers who get scholarships and the like that allow them to study longer, suggesting a positive relationship between talent and the length of study, and hence between earnings and the length of study.

saving; the stream of income in the form of interest over the duration of the deposit or the lifetime of a bond is the return on the amount that was deposited or purchased. Say you deposit £100 for two years and the rate of interest is 10 per cent, after Year 1 you earn £10 in interest and after Year 2, if you leave it in the account, you would earn £11 because of compounding: so your total return is £21 on an outlay of £100 over two years and the rate of return is 10 per cent. This simple example serves to show that it is necessary to take account of the compound effects over the stream of future earnings and bring it into the present. That is called discounting. What we need to do, therefore is to add up total earnings over a life-time and discount earnings back to a particular point in time, and then calculate the rate of return over the total cost of the training at the time at which they are incurred. But since costs are also incurred over a period of time they also need to be discounted, and this is done by finding the present value of costs and earnings. We shall follow this process through its various stages. The first stage is to establish age-earnings profiles and to use them to picture the relation of the stream of costs to the stream of returns over a working lifetime.

Age-Earnings Profiles

In Chapter 5, it was shown that earnings increased with age up to a certain point and then fell; plotting earnings by age on a graph produces an age-earnings profile. The profile of earnings begins at the age at which the person starts work, or at the average age of starting work in a particular profession, and continues until retirement. But before the age of starting to earn, expenditure has been incurred on training. The inclusion of training costs is illustrated in Figure 6.1. In Figure 6.1 a notional singer enters college at the age of 18 and leaves at 22; the costs of training occur between 18 and 22, rising somewhat during these years. At 22 the singer begins work and earnings begin to rise and reach their peak between 40 and 50 but then begin to fall; at 55 let us suppose that the singer gets a teaching job which he or she does until retirement. Costs incurred during training are shown below the earnings axis, being represented as negative earnings. Different situations can be pictured showing two different examples of training costs and earnings profiles; in Figure 6.2 there are two different cost situations

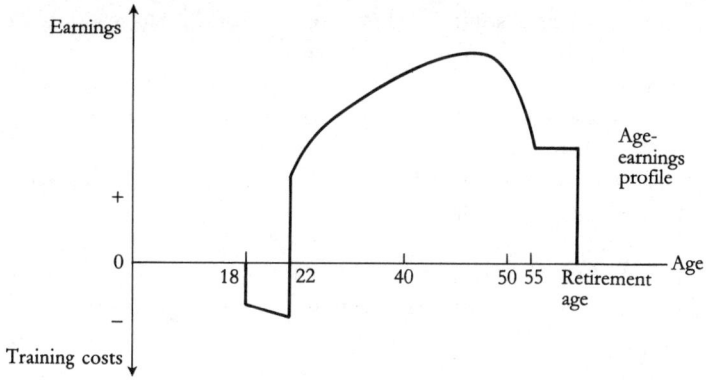

Fig. 6.1. A Typical Age-Earnings Profile

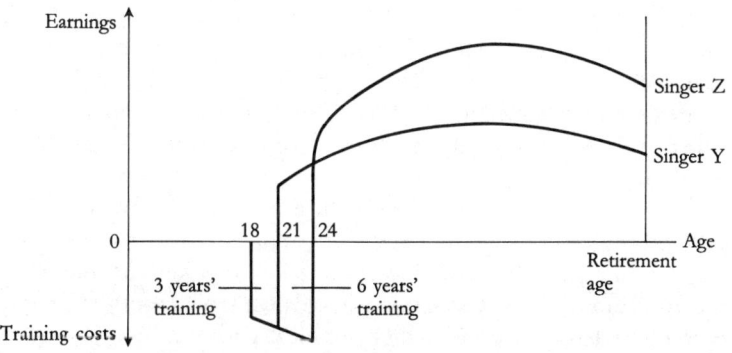

Fig. 6.2. Comparison of Two Possible Age-Earnings Profiles

illustrated and two different types of career patterns. Take two imaginary singers, Y and Z. Suppose that Singer Z trained for six years and had a financially successful career; Singer Y trained for three years and earned less at every age. Both started to study at the age of 18. But Singer Z incurred much higher training costs than Singer Y and so must earn more over a working life than Singer Y to recoup these costs. Even if they both earned the same at age 24, Singer Z has three more years of both direct and indirect training costs to make up for.

The longer that training lasts the more it costs in total and the later earnings are postponed. So what has to be done is to calculate how much more a singer with a longer training needs to earn to recoup the extra training costs. A formula is needed that allows us to calculate the value of the earnings stream over the whole working life of a singer and relate that to the total cost of the initial investment in training. This calculation can be made by finding the present value for the investment. The formulae and calculations are to be found in Appendix 2.

The calculation necessitates use of the notion of opportunity cost, for the total cost of training is not only the direct cost but also the indirect cost of earnings forgone as well, and the indirect costs of singers persist after the end of the formal training period in the form of lower earnings while the singer continues training in the form of on-the-job training. It is precisely this type of cost that is embedded in the rate-of-return calculation because the rate of return in fact measures the difference that additional education or training makes to earnings by virtue of the extra investment in human capital. Thus what is measured is the economic value of undertaking higher education by comparison to what earnings would be if the higher education had not been undertaken; the economic value of a degree or diploma is the addition to earnings that is due to additional education. In order to make this comparison it is necessary to try to standardize 'before' and 'after' characteristics, particularly educational qualifications, but also other factors that affect earnings, such as social class, innate ability, and the like. What we are doing in effect is looking for a 'control' group of people, who could have undertaken higher education but did not, and who share the same non-educational characteristics. Although it is not possible here to standardize for the latter, it is possible to compare singers' earnings with those of men and women who left school at 18 but went to work instead of going on to higher education.

This can be illustrated in a diagram showing the comparison of the direct costs of higher education and the indirect costs of forgone earnings. Figure 6.3 shows the age-earnings profile of an 18-year-old-school-leaver and the age-earnings profile of a graduate. The direct costs of formal higher education which take place from 18 to the age of graduation are shown below the axis and the amount of earnings forgone is shown by the sum of the 18-year-old school-leaver's earnings during the graduate's period of study, i.e. from 18 to the age of graduation. The difference that higher education makes to lifetime earnings is the area between the age-earnings profile of

THE RATE OF RETURN TO TRAINING

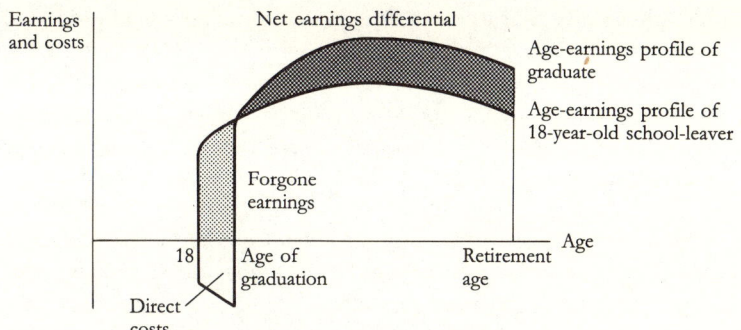

Fig. 6.3. Net Earnings Differentials due to Higher Education

the 18-year-old leaver and the graduate and this is called the net earnings differential (the cross-hatched area in Figure 6.3). The exercise, then, is to compare the present value of the net earnings differential with the present value of forgone earnings (the indirect costs: the hatched area) and the direct costs of training.

With information on the earnings of graduates and 18-year-old school-leavers it is possible to compute the indirect costs of higher education and the benefit of the investment in higher education, i.e. the net earnings differential by notionally bringing these back to the age of 18, the point at which the investment decision is made when the student decides whether to undertake higher education or go out to work. This is achieved by using the rate of return formula (see Appendix 2). The result will be the rate of return to higher education of one type or another, which can then be compared to the rate of return to different types of investment in human capital or to other types of investment. Within the field of higher education, rates of return have been calculated for degrees in different subjects.

Private and Social Rates of Return

In Britain, higher education is for the most part funded by state subsidy and therefore the direct costs of formal study, i.e. the cost of the place plus the fees do not have to be borne by the individual student but by taxpayers. Where the only costs that the individual has to bear are just forgone earnings, the private rate of return is the relevant factor in what is a private decision as to whether or not to undertake higher education. In this situation the individual is deemed to compare the private rate of return with the return that can be had from alternative private investment opportunities, such as the rate of interest on bank or building society deposits or government bonds. If for some reason a student has to pay the full fee covering the direct cost of the degree course, as overseas students do, then the direct cost must be added into the calculation of total costs.

The direct cost of formal training must be taken into account in calculating the social rate of return to higher education because higher education imposes resource costs on society. The cost of the place in a university, music college, or other institution of higher education is a cost to society because it is a subsidy to the institution paid out of general taxation. So the social rate of return represents the return to society of its investment in human capital formation through education. In calculating the social rate of return, earnings differentials are compared with the total resource costs; opportunity costs still feature in this calculation but they take on a different dimension—in the private calculus they represent earnings not enjoyed by the individual but in relation to the social rate of return they represent output forgone by society as a whole.[2] In finding the social rate of return, the social cost benefit ratio is what is being calculated and this is then compared to the return to other forms of social or public investment, such as road-building, health expenditures, defence expenditures and so on. The government has a rate of return which it uses to assess the viability of investment projects under consideration; this is the Test Discount Rate and the actual figure for this was between 5 and 7 per cent in the 1980s, depending on the degree of uncertainty associated with the costs and benefits of the project.

Both private and social rates of return to many types of education have been calculated over the last twenty-five years by economists. These are for education of all levels in various countries and include calculations of rates of return in different occupations, for example, lawyers, engineers and teachers. A study by Clark and Tarsh (1987) of the Department of Education and Science in Britain calculated that the private rate of return to higher education, taken over all subjects from 1979 to 1984 was over 20 per cent and the

---

[2] For a detailed explanation of this and related topics in this section, see Blaug (1970).

social rate of return varied from 12 per cent to below zero, depending on the subject area.[3] (The private rate of return is nearly always higher than the social rate because higher education is subsidized.)

Rate-of-return analysis in education makes a number of assumptions and is a far from perfect device for assessing how economically worth while particular forms of education are. The basic underlying assumption of the technique is that the benefits either to the individual or to society are measurable by monetary (pecuniary) earnings and that all earnings differentials standardized by age and sex are due to additional education rather than other factors (in technical terms an education or alpha factor of one). Studies have been done which attempt to remove the contribution of other factors, such as social class or race and innate ability from the return to education itself. It may be that in singing it would be impossible to remove the effect of natural or innate ability; if the only singers who were admitted to singing courses in higher education were ones who were so gifted that they would have made a career even without going to college, for example, then what is in fact being measured is innate ability or talent rather than the contribution made by their college training. Reference has already been made to the fact that in studies of artists' earnings the correlation between the length of study in formal education and earnings is weak. There are a host of objections: however, these objections apply to all rate-of-return studies to a greater or lesser extent and it is at least interesting to see how training in singing compares with other types of higher education.[4] It is perhaps worth pointing out here that if the government ever implemented a full student loan scheme coupled with charging full-cost fees for higher education, every student would have to calculate the private rate of return in his or her chosen field. How would singers fare?

Notional Rates of Return to Training Singers

I have calculated notional private and social rates of return for singers. It is important to be clear from the start about the results

---

[3] First degrees compared with all A-levels for men, including the maintenance award and assuming an education factor of 100% (alpha = 1) with an earnings growth of 2%. See Blaug (1970) for clarification.

[4] McMahon (1987), also Blaug (1987: 129–40) review the arguments for and against rate-of-return analysis.

that are presented below. These are not actual rates of return because the absence of statistically significant earnings data means that it has not been possible to do an accurate calculation. What has been done instead is to use the data on singers' earnings to make an informed guess about what a typical singer's earnings might be. I have used the term 'notional' to make that clear. The data on forgone earnings, the indirect costs of formal and on-the-job training are actual data published by the Department of Employment. The costs of formal training are also actual data taken from the Gowrie Report (Gowrie 1990); the figures are for the cost of a place in a music college plus the fee payment, assumed to be made by the student's local authority. The notional singer is assumed to have had a student maintenance award. All the figures used are for one particular year, 1989. The calculations are presented in Appendix 2. The reader is urged to try to follow how they are made, in principle at least.

Solving the equations given in Appendix 2 results in a private rate of return to training in singing of 4 per cent and a social rate of return of between 1.5 and 2 per cent. These rates of return may be compared with those calculated by Clark and Tarsh (1987) for various types of male university graduates in the 1980s. In Arts subjects the private rate of return was 10 per cent and the social rate of return was negative. The comparison is not really a reasonable one, however, since Clark and Tarsh used actual figures for earnings whereas the ones used here are notional ones. Furthermore, the rate of return to a music student in a university would come under the heading of Arts and it is certainly to be expected that the private rate of return to studying singing in a university would be higher than that of training in a music college because the training period in a university is shorter and the cost of a place lower. (Note that this calculation of the social rate of return would also be the relevant calculation for the private rate of return for students who do not receive a grant or if the government were to introduce a full student loans scheme to replace grants.) Given all the limitations of this exercise, it is nevertheless interesting that the social rate of return to the notional singer is positive, i.e. it is greater than the negative rate calculated by Clark and Tarsh. On the other hand, very favourable assumptions were made about the notional singer's career. But again, too much should not be read into these results.

One purpose in calculating the private rate of return in singing is to allow a would-be singer to make a rational economic decision about entering the singing profession. By looking at Table 21 in Appendix 1 a singer should be able to assess how economically worth while his or her training has been. Clearly, if a singer earns or expects to earn less than the notional earnings figures given here, his or her private rate of return would be lower. Equally, the longer the training, the higher will be the indirect or opportunity costs and the direct training costs, and the private rate of return would again be lower than 4 per cent. The earnings figures used to calculate forgone earnings in the rate-of-return calculation are for 18-year-old school-leavers, with one A-level: someone with two or more A-levels would earn more (see Appendix 1, Table 20), and so a singer with two A-levels would have higher indirect costs of training; again the private rate of return of 4 per cent would be too high in such a case. In fact, any one with a scientific calculator can work out their own private rate of return by drawing up a table of earnings and costs and using Formula 1 in Appendix 2; by choosing test values for r, the rate of return can be calculated.

The private rate of return of 4 per cent was below the market rate of interest that could be earned from an investment such as Building Society shares in 1989. That was about 10 per cent. So, it may be concluded that, using these notional earnings figures, the outlay on training as a singer was not a good financial investment. Alternatively, another type of calculation can be made, which is to estimate what lifetime earnings as a singer would have to be in order to cover the direct and indirect costs of training at a 10 per cent interest rate. This calculation is also shown in Appendix 2. It is a simple calculation of the kind that anyone with a savings account makes all the time! Table 22 (Appendix 1) shows that by the age of 35, a singer would have to earn £17,386 a year at a discount rate of 10 per cent to equate the stream of costs with the stream of earnings. Earning that amount is the equivalent of making 10 per cent interest on a lump sum investment equal to the amount of the outlay of four years' training at a music college. Table 22 also shows what singers would have to earn each year from the age of 22 to 35 to equate them to an Arts graduate whose private rate of return is 10 per cent. Could singers earn these amounts? Yes, successful singers undoubtedly could but it seems very unlikely that such earnings were achieved by the majority of people in the singing profession.

These calculations are based on the average earnings of 18-year-old male school-leavers with one A-level in 1987. By mid-1989 these would have increased by 23 per cent in line with the Average Earnings Index. Thus the figures in Table 21 would have to be increased by 23 per cent to make them comparable with the earnings data for singers presented in Chapter 5. That suggests that, for example, a singer who immediately went to work in an opera chorus on leaving college would not earn an income comparable to an Arts graduate of a university or enough to keep pace with the earnings implied by a a discount rate of 10 per cent. However, results are likely to be very different for male and female singers. It has not been possible to calculate rates of return based on female earnings because data are not available from the General Household Survey. As average earnings for non-manual females are always lower than male earnings, that suggests that the rate of return to women singers is higher than that for men, something that is always found in rate-of-return studies.

Risk and Uncertainty and Singers' Earnings

In these calculations it was optimistically assumed that the notional singer worked every year up to retirement age. They did not take account of the risk and uncertainty inherent in the singing profession. This takes the form of uncertainty about employment prospects, about the number of years that a singer can hope to sustain a demanding career vocally and physically, and even the risk of whether it will be possible to make any sort of career at all.

In economic theory, a distinction is made between risk and uncertainty, 'risk' being the term applicable to situations in which the probability of particular outcome can be calculated and 'uncertainty' applying to situations in which it is impossible to calculate the chances of some event happening. Average probabilities can be calculated in a variety of situations, indeed the whole insurance business rests on such calculations; without them no insurance company could be sure of covering its commitments or make a profit. The average probability of fire, theft, accidents, dying before a particular age, and so on can be calculated for the population at large, and the insurance company in effect takes a gamble knowing what the chances are in general of its customers making a claim. But in situations where no such calculations are possible, only subjective estimates can be made in the face of uncertainty.

In most occupations information can, at least in principle, be obtained about variables such as unemployment and the normal career span, which enable entrants to those occupations to get some idea about what the chances of success will be. In many occupations, for example, it is reasonable to suppose that employment will continue up to a set retirement age, such as 65 for men and 60 for women. In such occupations there is a calculable risk of unemployment and the individual can form reasonable estimates of his or her expected lifetime earnings. In the singing profession, however, such estimates are impossible and the individual's employment situation is uncertain.

Obviously, risk and uncertainty pervade many areas of economic life, particularly where investment is concerned because this involves making assessments of future returns. In capital markets, risk can be discounted by adding a premium to the expected rate of return, thus adjusting expected returns for variations in the future. The longer that an investment or capital project takes to mature, the greater the risk involved and the higher the rate of interest has to be to compensate for it. This is why riskier investments pay a higher rate of interest. Investors choose between investment opportunities, expecting a higher rate of return the riskier the investment and the further into the future repayments are to be made. People act differently with respect to risk, some preferring present certainty and others preferring to take risks; the former group are called risk-averters—people who shy away from risk—and the latter group risk-takers. This distinction applies to occupational choice as much as to investment decisions and this is not surprising, since the decision to invest in human capital in the form of education and training is an investment decision. Therefore, people who enter occupations in which there is considerable variation of earnings and uncertainty about future prospects are assumed to be risk-takers.

Risky occupations, i.e. those in which there is a small chance of earning very high incomes and a great chance of low earnings, are clearly ones in which the distribution of earnings will be skewed in the way described in Chapter 5. Indeed, as Rosen (1981) remarks, the prevalence of risk may provide an alternative explanation to his own of the skewness of earnings distributions. Use of the term risk, however, implies that probabilities can be assigned to the chance of success or failure. Choice is viewed as being akin to entering a lottery in which the structure of prizes is known and the chance of winning a prize can be calculated. Though the individual cannot know if he or she will be lucky, nevertheless the probability of one person being successful can be worked out, at least after the draw has taken place. How much people pay to enter the lottery is determined by the chance of winning and successive events will reveal what this amount should be (just as the probability of rolling a six on a die can be calculated by experiments in which dice are rolled many, many times). But this type of situation is very far from being applicable to the decision to enter a profession such as singing, where the chance of success cannot be estimated objectively.

But just because a singer's chance of success cannot be estimated objectively and the actual degree of risk calculated, this does not mean that the analysis of risk is irrelevant. Singers do form subjective probability estimates about their chances of success—indeed, they almost all believe at the outset that they will rise to the top of the profession—and they do have an idea about the size of the prize in the lottery that is the singing profession, because they hear of fabulous fees paid to top singers, even if hearsay reports are ill-founded—Pavarotti getting £100,000 for a concert for example. Adam Smith aptly observed in 1776 'the contempt of risk and the presumptuous hope of success are in no period of life more active than at the age at which young people chuse their professions' (Smith 1976: 126), pointing out that the chance of gain is naturally overvalued while the chance of loss is frequently undervalued. Smith had recognized even as early as 1776 that earnings in different employments would vary according to 'the probability or improbability of employment in them.... In a perfectly fair lottery, those who draw the prizes ought to gain all that is lost by those who draw the blanks. In a profession where twenty fail for one that succeeds, that one ought to gain all that should have been gained by the unsuccessful twenty' (Smith 1976: 122–3).

This analysis explains two aspects of the singing profession: first that young singers' subjective assessment of their chances of success are inevitably too high, and secondly, why earnings are unevenly distributed. Even if young entrants to the singing profession knew that the average private rate of return was low or even negative, they would nevertheless hope to do better than the average and believe that they have a good chance of winning the highest prize by reaching the top of the profession where high earnings and public acclaim are the rewards. All singers implicitly recognize the analogy of the lottery—that luck has as much a role to play in a career as talent; they know that they are risk-takers. The absence of objective information on the degree of risk, the fact that the singing profession is an uncertain one, serves to exacerbate the tendency to overvalue one's chances. The whole process of learning to sing and perform is almost impossible to contemplate without a total belief in one's own ability; as one teacher said, 'without the fantasy of being great you could not even begin'. In the individual case it is perhaps impossible to get an objective assessment; however objective singing-teachers may try to be, one-to-one teaching naturally involves the teacher with the pupil in a way that makes objectivity difficult to achieve. This leads to unrealistic assessments of the chances of success. One of the most beneficial aspects of attending a large music college and entering competitions is that students become aware of others' abilities and can match themselves up to them. Self-assessment is an important part of the information problem of singers!

Furthermore, as a singer makes progress in a professional career, the odds against failure shorten as objective evidence of the chance of success is produced by the degree of success so far attained. More information about one's chances is, as it were, made available. Once a singer has established his or her ability to get professional work, the future would seem more assured, even though uncertainties in fact still persist about how long the success can be maintained and the unknown factors such as vocal health and how long the career can be sustained still cannot properly be evaluated. Clearly, however, the chances of success in the long run increase with the extent of the singer's record of success; thus the subjective probability assessment depends on the stage that a singer has reached on his or her career path. Once a singer has reached the top of the profession, the risk of failure falls almost to zero. This is consistent with MacDonald's explanation of why earnings distributions in professions such as singing are skewed, which was outlined in Chapter 5. His analysis, however, does not take account of the role that risk plays in the information problem facing singers and other such professionals. The question of trying to assess risk and uncertainty is basic to the information problem that singers face; if they had the information necessary to make a proper assessment of risk and could estimate their future earnings and employment prospects objectively, they could make an informed decision about what rate of return on their investment in training they expect to obtain, that is, they could discount for risk in advance of undertaking training rather than learning the hard way. But without that information they can only form subjective views about their chances of success and they are likely to overvalue their abilities and chances. One of the chief problems about information in the singing profession is that everyone knows about the success of the stars but nothing is heard of the many failures; this, again, causes people to overvalue future earnings.

Is there any way that the probability of financial success in the singing profession in Britain can be calculated? Perhaps the following could give a very rough indication, though it is almost pure guesswork. We know that there were around 350 graduates and diplomates of courses in all types of institutions of higher education which offer singing as part of the final examination in 1988/9. Assuming that student numbers are stable and that people fully enter the singing profession at the age of 22 and retire on average at 60, the working life of a singer is thirty-eight years. This gives a stock of $350 \times 38 = 13,300$ singers in Britain who have received higher education (this ignores people training privately). Let us assume that as many as 50 per cent of these are lost to the singing profession through changing to other occupations, marriage, death, etc; then we have a stock of 6,650 working singers. Let us now guess that 66 of these are highly paid artists and 665 work regularly making a good living; these represent 1 and 10 per cent respectively of the total number of singers. Of course, we have no way of knowing whether or not this is a reasonable guess—but if you ask heads of singing departments how many of their students they expect to make a career, they will tell you two or three a year—maybe. However unrealistic these figures are, it can safely be said that only people who have a strong preference for taking risks should enter the singing profession.

What can be predicted from economic theory is that in professions in which risk-preference predominates, there will be overcrowding or excess supply, since all entrants overestimate their chances of success. Adam Smith, who was quoted earlier, recognized this and attributed the lure of such professions to the psychic benefits of success: 'To excel in any profession, in which but few arrive at mediocrity, is the most decisive mark of what is called genius or superior talents. The publick admiration which attends upon such distinguished abilities makes always a part of their reward; a greater or smaller in proportion as it is higher or lower in degree. It makes a considerable part of that reward in the profession of physick; ... in poetry and philosophy it makes almost the whole' (Smith 1976: 123). Hence, by taking account of subjective attitudes to risk and uncertainty, not only can we better understand individual decisions in relation to the expected private rate of return in a profession such as singing, but we can also explain the aggregate effect on supply and predict that overcrowding will result. Furthermore, the fact that there is an excess supply of entrants depresses average earnings in the profession, and this both increases the gap between the highest earnings and average or median earnings and also increases the earnings differential between singing and other occupations as well, further lowering the private rate of return. There is a depressing dynamic to this situation because as the earnings of top stars increase (the prizes in the lottery get higher), the profession becomes more attractive to people who have a preference for risk and there is a further downward spiral of average earnings and the rate of return. I put this forward as a much better explanation of skewed distribution of earnings than those of Rosen (1981) or MacDonald (1988), which were discussed in Chapter 5. One way out of this situation is surely better information about earnings and employment (or, rather lack of earnings and unemployment) in artistic and other such professions so that would-be entrants understand the risk that they are taking.

Ignorance of the role that risk-preference as opposed to the presence of risk, plays in understanding the decision to enter artistic professions is well illustrated in a study on uncertainty and investment in human capital in the arts by Waits and McNertney (1980). They defined risk in arts professions in terms of the greater variability of artistic earnings compared to earnings in other professions in the United States. Specifically, they analysed the variation from average (mean) income in arts professions and compared this measure of the distribution of arts incomes to those in a selection of other professional occupations using US Census data on incomes. By showing that there is greater variability in artists' earnings than in those of other professionals, they concluded that artistic professions are riskier and deduced from this that the rate of return to human capital investment would be lower, resulting in underinvestment in arts professions, i.e. fewer entrants than is socially desirable. Their overall conclusion was that this situation should be corrected by subsidizing students to undertake training that will lead to a career in the arts. (There is no system of government subsidy to higher education in the USA.) This conclusion, perhaps argued differently, is frequently held by many people in the arts. It is implicit in UNESCO-thinking about the Status of the Artist and has supporters all over the world. I believe it is not only misconceived but counterproductive to the interests of artists.

The argument, however, fails to distinguish private and social rates of return, but more significantly, it completely fails to take account of risk-preference, which, as I have argued, not only characterizes artistic professions but acts as a spur to entry; it is precisely because there is variability in earnings—the result of some people having very high earnings—that risk-preferring individuals are encouraged to try their luck. Risk-preference causes overinvestment not underinvestment in human capital in the arts, resulting in an oversupply of trained artists rather than an undersupply. Surely, it must therefore be concluded that there is too much subsidy rather than not enough.

A much more satisfactory approach to the question of risk and uncertainty in the performing arts was taken by Santos (1976), again using US Census data on earnings. Using a present value estimation formula she attempted to calculate not the rate of return, but the net earnings differential that would be needed to compensate dancers and singers for the extra training that they undertook in comparison to workers in other occupations. Though Santos herself did not consider her results to be reliable (and there are a number of quite incorrect assumptions made about the costs of training singers), she showed that the compensating differential was too low, implying a negative private rate of return to the investment in human capital. Santos, by contrast to Waits and McNertney, concluded that the larger variance in the earnings of singers and dancers compared to those in other occupations served as an attraction to prospective entrants, concluding that 'risk-preference and psychic income apparently prevail over financial considerations when considering the pursuit of a career in the performing arts' (Santos 1976: 257).

Where risk is prevalent in capital investments, the rate of return is raised to take it into account. Instead of a rate of return of 10 per cent that could be obtained in a safe Building Society account, a premium must be added for risk. The Treasury Test Discount Rate added an extra 2 per cent to allow for risk in calculating the return on government projects the 1980s. This diminishes even more the notional private rate of return of 4 per cent calculated above.

Rate-of-Return Analysis and the Planning of Higher Education

Overall provision of higher education in Britain is planned by the government. As higher education is subsidized, the chief medium of planning is the control of the amount of subsidy out of general taxation that is allocated to the sector. The government also sets overall target student numbers and issues directives which affect student numbers. These directives concern the split between science and technology places in higher education (45 per cent of places are reserved for student places in science and technology courses), and for medicine, and upon occasion teacher-training, the government sets specific subject target student numbers. It also determines the fee that institutions of higher education can charge for courses. However, it leaves the details of the distribution of subsidy and questions of target student numbers on courses to the two bodies to which it allocates funds for distribution to individual institutions; in 1990 these were the UFC and the PCFC.

These bodies had their own criteria for allocating funds to universities, polytechnics, and Colleges of Higher Education, and also to the music colleges. The institutions themselves are autonomous in the sense that they control their own courses and standards and they are to some extent free to decide the number of students they admit and how they spend their resources. But in the late 1970s, and even more during the early 1980s, the UFC and the NAB, the forerunner of the PCFC, adopted a more interventionist policy and sought to control target student numbers in individual institutions. Again, they did this by a mixture of directives and control of the purse-strings, using the unit of resource per student that they allocated to agreed student numbers in broad categories of courses. Institutions of higher education were free to ignore UFC and PCFC directives but they suffered financially if they did so. The UFC in particular, but also the PCFC, tried to protect the unit of resource and this restricted the supply of places in universities in relation to the demand for them. In 1989 the government increased the student fee in order to offer incentives to institutions of higher education to expand student places but there were still more qualified would-be entrants than places and these places were mostly allocated by A-level grades, places going to those with the best A-level performance.

To say that students had a free choice as to what courses they studied, then, meant that if they had good enough A-levels they could pick and choose. Economists see individual choice as being guided by private rates of return of different types of degree subjects. In the 1980s the private rate of return was highest in engineering (32 per cent) with Social Science subjects coming second (27 per cent).[5] However, students do not necessarily aim for the highest rate of return because they are motivated by other factors, such as psychic benefits, rather than by purely financial considerations. This is clearly the case in singing.

Just as the private rate of return measures the individual's opportunity cost, so the social rate of return is a measure, albeit an imperfect one, of the social opportunity cost to society as a whole of deploying its resources one way rather than another. Because all resources have competing uses and are in scarce supply in relation to the demands on them, governments have to decide how to spend their resources in such a way as to achieve social goals. Education competes with health, defence, road-building, etc. for society's resources, and within the education sector, schools compete with universities and other subsidized institutions. Should the government allocate its resources to those which yield the highest social rate of return? Many economists believe they should, even though it is inherently difficult to measure the social benefits of particular undertakings.

The government does not use social rates of return in deciding how much to spend on health rather than education, though it does use them in evaluating specific projects, such as building a hospital or school; it then compares the social rate of return of the project to its Test Discount Rate and may reject projects that fail to equal or exceed it. Nor does the government use social rates of return in planning higher education and neither did the UFC or the PCFC use them in deciding how much to allocate to particular subjects. So, at the time of writing at least, the social rate of return had only an academic interest and did not form the basis for planning in higher education. Nevertheless, the DES has measured rates of return

---

[5] Clark and Tarsh (1987) calculated on the same basis as explained in n. 3, above.

over broad subject categories and, since it has never made clear the basis on which it did plan higher education, it might be influenced by them to a limited extent.

The Economic Case for Subsidy to Education and the Arts

Why do economists believe that rates of return should be used to guide decisions about government spending on education and in other sectors of the economy which receive subsidy, such as the arts?

In practically every country in the world, education is subsidized out of taxation and in many countries of the world the arts are also subsidized by governments, indirectly if not directly. In Britain expenditure out of taxation (as opposed to support by private patronage or charitable foundations) on education began in the nineteenth century, but central government subsidy to the arts has been a relatively recent phenomenon; though there was some small piecemeal public subsidy during the first half of this century, the pattern of arts subsidy to the performing and visual arts and the heritage was essentially established post-1945. However, it is worth while remembering that the possibility of fully subsidized (i.e. 'free') higher education for those with adequate qualifications was also a post-war phenomenon.

What is the economic argument for this subsidy? The economic case for subsidy rests on the idea of 'market failure'; moreover, the arguments for subsidy to education and the arts also provide the logic behind the advocacy of cost benefit or cost effectiveness analysis, in which the social rate of return is the variable that should be used to guide government decisions about what and how much to subsidize.

As a prelude it is necessary to understand how economists approach the question of subsidy. Belief in the efficacy of free markets is fundamental to economic theory. Where markets can work freely to determine the prices of goods and services and the quantities to be produced, it is usually argued that they should be left to do so. Markets work through supply and demand; supply is determined by the costs of production of the resources used to make the good or produce the service and demand is determined by consumers' willingness and ability to buy the goods and services. Consumers are held to be the best judge of their own tastes and preferences and purchase quantities of different goods and services so as to satisfy their needs and wants; this is the doctrine of consumer sovereignty. Producers are motivated by profit to provide what consumers demand, but because they have to compete with each other for resources and to attract custom this results in goods and services being produced at the lowest possible cost. Therefore, competition, both between consumers and producers, results in efficient prices because costs are forced to the minimum possible to cover the use of resources. Price is the guide to decision-making in the market economy. The free market also determines wages that reflect the interaction of supply and demand for labour and the same applies to other factors of production; the wage rate, the price of a unit of labour time, reflects the value that labour adds to the productive process. Thus the market system simultaneously solves the problem of how much to produce, the distribution of the goods and services produced, the payment for the use of resources and the distribution of income.

But the market system does not always work in this ideal way. Where there are externalities the market mechanism does not work properly. Externalities are costs or benefits that are not taken into account in prices; if what one individual produces or consumes affects the costs or benefits of another, imposing higher costs or benefits on that individual so that the price he or she paid does not represent the true resource cost or benefit, external effects are present. If I buy beautiful trees and plants for my front garden, which I do for my own pleasure, I benefit people who live nearby and passers-by who see it for nothing. Because they enjoy my garden, they may be willing to pay me to keep it beautiful but there is no price mechanism for so doing.

Even if I put a box for contributions on the front gate, no-one is obliged to pay anything and they can enjoy their view of my garden for nothing. Thus my desire for a beautiful garden produces both private benefit (mine) and social benefits (to neighbours and passers-by) though only I pay the cost. By contrast, if I care nothing for my front garden or what anyone thinks and use it as a tip, I impose costs on other people in the form of ugliness, smells, and possibly health hazards, the external effects of my choice about my garden. Social benefits are defined as being positive externalities and social costs are negative externalities.

Where externalities are present, the state may have to intervene in the working of the market by regulating it so that prices are made to reflect the true cost of resources.

One way that this can be achieved is through subsidies and taxes. Subsidies are used to compensate producers for social or external benefits that the market price does not cover. They cause producers to produce more at the market price than they would otherwise do. Subsidies are therefore used when the market cannot properly price goods and services and when society wants to encourage people to do things that are in the interests of everyone and not just in their own interest. Subsidies work through economic variables, usually by lowering prices and raising the revenues of producers. Of course, there are many subsidies that seem to be contradictory in purpose. The EC subsidizes farmers to produce more milk, wine, and beef than people in the member countries want to buy, at least at the prices that are charged (and these prices are in turn regulated by the EC to protect farmers). The result is butter and beef mountains and wine lakes. What this suggests is that subsidies must have a clear policy objective as well as be efficiently operated.

The economic argument for subsidies to education and the arts rests on the presence of external benefits. These benefits are perceived as being of a general nature; both education and the arts have a civilising effect on the population at large, being instrumental in raising the quality of a society's political and social life through their effect on social cohesion, improving the ability of the population to make reasoned decisions (such as voting), respect law and order, and form common social goals. To function properly, a society must be literate, creative, and articulate. Education and the arts could be provided entirely by private enterprise, since it is possible to have private schooling and commercial theatre, opera, dance, art galleries, and so on, but providing these goods and services through the private market will result in a level of output that is below what is socially desirable because of externalities. The presence of these external benefits means that the market price can never compensate producers for all the benefits to society and hence there will be underprovision, unless public subsidy encourages producers to produce more and consumers to consume more. But the question is whether the social, as opposed to the private benefits are sufficient to merit the outlay in subsidy, i.e. the cost of social provision. If the subsidy is greater then the social benefit the social rate of return would be negative. The extent to which social benefits exceed the amount of the subsidy is measured by a positive social rate of return so the social rate of return measures how efficient the subsidy is.

But it is not only a question of social efficiency, there is also the question of how equitable subsidy is. It has long been recognized that there is a tendency for better-off members of society to benefit more from public subsidy in many areas of activity—middle-class children benefit more from subsidies to education, particularly those to higher education, since entry to higher education is still dominated by middle-class students despite years of public subsidy: audiences for the arts tend to be drawn from the better off and more highly educated sections of society. Yet taxes are paid by all income-earners and the bulk of taxation is financed by poorer people simply because there are more of them, and tax rates are not sufficiently progressive to counterbalance this. Therefore subsidies to higher education and the arts tend to redistribute income from poorer to richer people rather than the other way around and this is inherently inequitable. These are difficult problems and they suggest that the case for subsidy must be carefully made.

Subsidies to Training Singers

What light does the above discussion throw on the question of subsidies to training singers? In Chapter 1 it was reported that there were approximately 200 students leaving music colleges with a qualification and that they spent at least four years in college. This suggests that there were over 800 singing students in music colleges in 1989, the 'over' making allowance for drop-outs and students who failed to qualify at the end of their courses and students who stay on for more than four years. With the annual cost of an undergraduate place in a music college in 1990 being around £6,400, this suggests that over £5 million was spent annually on subsidizing places for singers in music colleges, to which figure must be added expenditure on student maintenance grants for those students who received them. The maximum grant was £2,500 in London, though of course, all students did not receive a grant, nor the maximum if they did. If all students had the full grant to study in London, that would add a further £2 million to the amount of the subsidy. In addition to singing students in music colleges, it was estimated that a further 140 or so students graduated from universities, polytechnics, and Colleges of Higher Education from courses in which singing was part of the final examination; they were in college for three or four years. With these courses costing approximately £3,300 a year (see Chapter 2) this added over £1.5 million to the public subsidy for the training of singers. Making the same assumptions as before about student grants, maintenance awards to these students would be over £1 million. These calculations suggest that the total public subsidy to training singers through formal higher education was in the region of £10 million. This is a very small percentage of the total higher education budget but, to put it in another perspective, represented about 5 per cent of the Arts Council grant-in-aid in 1990.[6]

There are other subsidies to training singers besides state expenditure on higher education. The National Opera Studio received a grant of £100,000 in 1990. A further source of public subsidy to singers after they have completed formal training was unemployment benefit and publicly funded schemes for the unemployed such as those of the Training Agency, Adult Education programmes, and suchlike; it is impossible to put a figure for subsidy to singers from these sources. Another source of subsidy, this time in the form of private patronage, was awards to singers from private foundations. Private foundations and trusts, etc. which give awards and scholarships and run competitions for singers gave over £300,000 to singers in 1988.[7]

Why does society support subsidies for training singers? After all, the social rate of return to their training was very low, according to the notional estimate made earlier in the chapter, showing that on financial grounds this was a poor investment for society to make. But the social rate of return could not take account of the

[6] Recurrent expenditure for universities, polytechnics, and colleges of higher education in 1989/90 was £2,700 million (not including student awards) so singers' training was about a 1/4 of 1% of the total.

[7] This figure comes from a questionnaire survey I undertook to all the organizations listed on the Arts Council's 1988 List of Music Competitions, Awards, and Scholarships and it was the value of monetary awards to singers in 1988 (or 1987 where the award was biennial). This figure would seem to underestimate the total value of awards. Some prizes included non-monetary benefits such as a Wigmore Hall recital, which was then worth well over £1,000; also a few organizations did not respond, thus underestimating the total. Furthermore, the figure did not include scholarships, etc. open only to a limited group, such as specific scholarships in music colleges. The true figure, then, was probably quite a lot higher. A total of 157 awards to singers were reported in 1988.

unmarketed benefits of training singers, since it was based on only the measurable benefits of private earnings; the presence of external benefits in education implies that the social rate of return underestimates the true social value of training and education (and a proportion of external benefits of the arts should also be added in too if only they could be measured). To be cost effective, these social benefits would have to be about £10 million, i.e. the total amount of public subsidy spent on training singers. But even if this figure were a true measure of the social benefits which a private market cannot take into account, a further problem has to be recognized. The purpose of subsidy where external benefits are present is to encourage more of the activity to take place than the private market would provide—in the case of training singers, to get more young people to train as singers than would do so if they had to pay for their own training. But the rate of return to singers was low because more singers were being trained than the labour-market for singers could absorb, while the cost of training for singers, being subsidized, must have encouraged more to enrol in higher education than would have been the case without subsidy. Therefore subsidies to training singers are part of the cause of the low rate of return. The fact that the average private rate of return is probably also low indicates that the singers themselves do not benefit financially from training; indeed, many who fail feel that they have been misled by society into training for something that cannot offer them a financially viable career—the hopes that subsidy raised by offering them the chance to train have been dashed by the cruelty of labour-market forces.

What this suggests is that the subsidy to training singers is too high because it encourages too many to enter the singing profession. This is not to say that there should be no public subsidy for the training of singers: after all, the rationale for subsidy is not only an economic one. Society has non-economic reasons for wanting to promote higher education; equality of opportunity is a social objective which has a wider appeal than economic arguments. Even with the present subsidies to training singers, concern has been expressed that only students from better-off families were able to go in for a career in singing. This is no doubt due to two of the factors that we have analysed, one being that it is a risky profession in which the chance of earning a living is relatively low, and the other is that there are much higher indirect costs involved in training to be a singer than for other graduates. Without subsidy this situation would worsen and reduce equality of opportunity for children of less well-off parents, which is socially undesirable. Furthermore, society believes that self-expression should be encouraged and the system of higher education is accordingly orientated towards satisfying students' tastes and preferences for courses. The expansion of places in higher education is demand-led and it is the stated aim of educational policy to encourage institutions of higher education to respond to student demand. That is the reason behind the development of courses in music, creative arts, drama, and suchlike, and there is every reason to believe that the number of places on these courses will expand as more students who want to study creative, self-expressive subjects enter higher education. That they do so for consumption rather than investment purposes is perhaps an unintended aspect of the policy but so far governments in Britain have not sought to restrict entry to particular courses (with the one or two exceptions mentioned earlier). Therefore, for reasons of social equity, singers have as much right to their subsidy as those studying any other discipline.

But if equity rather than social efficiency is the goal of subsidy, is the social rate of return a relevant consideration? It is clear that it would not be the only variable taken into account but most economists would nevertheless advocate that it be considered. What rates of return indicate is whether a project or policy is worth undertaking on financial grounds. The notion of the private rate of return is easy to accept, and society does not prevent people from spending their resources (time and money) on projects that have a low or negative rate of return, because of the belief in freedom of choice. If a singer with few prospects of success pays for his or her own training, knowing that the chance of making a positive return on that investment is low, one would assume either that he or she had a tremendous preference for risk or that the psychic benefits or consumption benefits were very high. However, things are different with regard to public spending out of taxpayers' money; ultimately public subsidy must be the result of social choice. Economics can aid that choice by placing a financial value on the policy. If a subsidy of £10 million is spent on training singers, that puts a 'price' on the policy; this price is the opportunity cost to society of other projects that cost £10 million (a hospital, another piece of military equipment). Ten million pounds, according to this way of thinking, would be the price of equality of opportunity for singers in Britain if the social rate of return were zero. But if the social rate of return is positive that indicates that there is also some financial benefit to society.

In this chapter the private and social rates of return to training in singing were calculated. They were necessarily hypothetical because they were based on notional earnings of singers. What has been achieved by calculating them is to show, given the known costs of training singers, how much singers would have to earn to make the private and social rates comparable to those of other types of higher education or other social programmes on which subsidy is spent. The use to which these calculations may be put has been argued on economic grounds, and has been shown to depend on which viewpoint is taken, that of the individual singer or society at large. Economic arguments are not the only ones that should be taken into account in deciding how individuals or society should spend resources, but they do serve to clarify the issues involved.

Conclusions

Does the Labour-Market for Singers Work?

The purpose of this book and the research that was undertaken in preparing it was to collect data about the economic aspects of the training and employment of singers in Britain and to apply economic theory to analyse the singing profession. This was done by looking at how the labour-market for singers works in general and by presenting facts and figures for the period 1988 to 1990 (with some going back to the 1970s) about the supply, demand, and payment of singers. One conclusion must be that this type of analysis can usefully be done, that, different though singers are in many respects from other workers, the labour-market for singers is not fundamentally different from other labour-markets; economic analysis, which has been developed and applied to the workings of labour-markets in general, has an explanatory power in respect of the singing profession—all that has been done here is to draw different strands of theory together and to juxtapose them, so only minor theoretical developments have been necessary. Despite the various institutional interventions in the labour-market for singers, such as minimum rates of pay, subsidy to the training of singers on the supply side, and subsidy to the arts on the demand side, the labour-market for singers works, though not perfectly.

To say that the market works is to say that a myriad of individual decisions are co-ordinated by the economic incentives provided via the price mechanism. These decisions include the decision to train as a singer, whether to make a career as a singer, how much work the professional singer is willing to undertake at various rates of pay, and what type of work to specialize in; these are all decisions that determine the supply of singers in aggregate and to particular branches of the singing profession. On the demand side, employers of all kinds who hire singers decide what works to promote and hence what type and how many singers they wish to hire; the outcome is determined by what they have to pay singers, how much they can charge for the performance—the price of a ticket or the record—and also by the proportion of the total costs that payments to singers constitute. Supply and demand determines rates of pay and therefore how much singers earn. The market does not work entirely freely, because wage rates and conditions of work are negotiated by collective bargaining between employers and Equity and the ISM representing singers but they only set minimum pay rates. Payments above the minimum are arranged by singers' agents according to what the market will bear. The resulting distribution of singers' earnings is very uneven with a few singers earning very high incomes and the majority earning considerably lower incomes.

The labour-market for singers is also affected by government policy. The fact that many of the organizations that employ singers are subsidized in part by government means that the demand for singers is to some extent dependent upon levels of subsidy to the arts (though the link between that demand and the amount of subsidy is weak, since only a certain proportion of the organizations' income is from subsidy and only a certain proportion of that is spent on singers). On the supply side, places in music colleges and in other institutions of higher education in which singers train are subsidized by the government and students are subsidized to study in them through student grants, which pay fees as well as maintenance. This subsidy undoubtedly encourages more students to study singing (and other subjects) than would be the case if subsidy were withdrawn. It is not known what the marginal effects of subsidy to higher education are, i.e. how many students would not go in for higher education if they had to pay more for it themselves. To that extent it is hard to quantify what the effect of subsidy is on the supply of singers; overall, however, it must encourage more to take up singing than would otherwise be the case.

But it is not only subsidized higher education institutions that train singers; there is also a well-developed private market for studying singing and it is reasonable to suppose that if subsidy to training in higher education were cut, there would still be a flow of supply to the market of singers training privately. Since there is no licensing or any such barrier to entry in the singing profession either for singing-teachers or for singers, singers who have trained privately stand as good a chance of getting professional work as college students do, provided that they have been as well trained.¹ For all these reasons, therefore, the labour-market for singers works despite the element of government influence on it via subsidy to the arts and to higher education. Supply and demand do work, albeit somewhat affected by the activity of government and the existence of minimum rates of pay.

But to say that the labour-market for singers works is not to say that the results are desirable. It is important to distinguish value-judgements from statements of fact or hypotheses. It is a fact that the singing profession is competitive and that only a few will make it to the top. The fact that there is competition in the labour-market, both between singers and between employers demanding them, is what makes it work; if this results in low earnings and unemployment as a consequence, we may deplore the situation but we have to recognize that it is a fact of life. Many people, particularly in the arts, have a great deal of animus against the operation of market forces and seek to modify or replace the market as a means of allocating resources and distributing incomes. They reject price as a measure of value, particularly where what is being priced is something as precious and intangible as artistic genius or the gift of a great talent, such as a singing voice. That is all well and good, but it is worth while considering what the alternatives are to leaving it to the market to solve the question of rates of pay and employment. Consider, for example, the full implications of adopting a planning approach.

What would be involved in organizing the training, employment, and payment of singers by administrative fiat? The flow of supply of singers could be controlled by restricting the numbers training in singing, say, in music colleges; so many places would be designated for training singers over a suitable fixed training period. To protect the designated flow into the profession some device would have to be used to prevent people from training elsewhere and claiming to be trained singers; that would most easily be done by

---

[1] The question of the quality of singing-teachers and methods of teaching singing, both of which are fraught with difficulties, has been avoided in the book. It is possible to study privately with highly acclaimed singing-teachers; most of the teachers who work in music colleges take private pupils and some highly acclaimed teachers only work privately. The real problem is that a teacher who is good for one pupil might be hopeless for another and standards are exceedingly hard to monitor. A further point is that training to be a singer involves much more than simply having singing lessons; that was made clear in Ch. 1.

having certificates that allow people to say they are trained singers (a licence, in effect) and allowing only music colleges to issue the certificates for the designated number of singers they are allowed to train. The number of trainees would have to be found through manpower planning; that would involve every employer of singers, opera companies large and small, choral societies, orchestras, choral conductors and fixers, and the rest in stating what their requirements would be over a future period, say, the next five to ten years, so that an estimate could be made of the number of singers they need. But it would not stop there; first of all they would need to specify the type of singer they need (voice type and range, etc.) which implies that repertoire is decided upon years in advance. That is certainly feasible, because opera companies and other promoters usually book singers several years in advance. Further, the authorities planning all this would have to decide what are normal hours of work for a singer and expect them to work a certain number of years as a normal career, perhaps from the ages of 25 to 55. Only then a figure could be produced for the amount of work available for singers and, given the numbers actually working in the profession (the stock of existing singers), the inflow of new entrants could be worked out and the number of places in music colleges and other institutions of higher education fixed accordingly. As newly trained singers come out of college ready to work, they would be fully employed so the problem would be simply to find out where the work is; some central register or agency could be set up to centralize information or singers could rely on job adverts in a professional publication. All this would have to be co-ordinated by some central office which would have to be set up to deal with this work.

What about rates of pay? After all, organizations employing singers do not just want singers at any price; their demand for singers to some extent depends on the fee they have to pay in relation to the prices they can charge for tickets, so they are not going to commit themselves to employing a certain number of singers over the next five to ten years without knowing the rate at which they would have to pay them. Employers and the singing profession would have to agree a standard rate of pay for every type of work which would then have to be binding on everyone; again, that is clearly possible—Equity has done this for many years. But it would also have to be set years in advance as well. There would be difficulties, such as singers going abroad to work who would have to be allowed for in the planning process; so that the manpower figures do not break down, there would have to be some sort of rules about working abroad and singers coming in from abroad would need a licence.

It is obvious from this scenario that replacing the market for singers by some sort of planning mechanism would be exceedingly complicated and would require the introduction of a large number of rules and regulations; it is possible but at a high cost. This sketch also makes it clear what is involved in the manpower planning approach to the question of how many singers should be trained.

One of the advantages of planning the training and employment of singers would be that the government could project exactly how much it has to spend on providing places for training singers in music colleges; and it could guarantee the amount of resources made available to student singers. Since the supply of singers would be restricted it would become vital that everyone trained, barring unforeseen accidents, would be able to go on to a career. Colleges would have to take great care in their selection of new students (something that is very difficult indeed, because it is very hard to anticipate the true development of singers at the age of 18 or even 20). There are no objective criteria of potential, such as A-level grades, which can be used as a guide. From the point of view of the singing profession (those actually in it, that is), the great advantage would be that they could expect to get employment and there would be much less uncertainty about their earnings. There could also be some kind of career ladder built in to the pay structure with 'promotion' based on merit, with scope for quicker advancement for the more talented singers, but this would be bureaucratically arranged rather than come about through the market's response to the relative shortage of talent. Nor would the distribution of income between singers be so unequal as it is when the market determines how much individuals work and get paid; a structure of wages and fees, even one paying differential rates, always produces a much lower spread of earnings than the free operation of the labour-market does. These are advantages that many people would like to see come about.

It is interesting to note that the resulting picture is not entirely unlike the situation in the Federal Republic of Germany (in 1990). Germany has always been held up as a shining example of a country that truly values the arts and artists; it spends much more per head in arts subsidy than Britain does, and there is little unemployment of singers because there is a great deal of work for them (see Appendix 3). But though the market for singers in Germany works more freely than the imaginary scenario sketched above, it is even so subject to considerable bureaucracy and one of its undesirable features appears to be a sort of 'Welfare State mentality'—singers have become somewhat entrenched in their ways, protected by bureaucratic regulations. Whether this is the result of a more bureaucratic system than that in Britain or some fault in the German training system it is hard to say; certainly many economists believe that bureaucratically planned economic systems lack flexibility and incentives for individual initiative, as well as being inefficient and costly to organize. There is certainly ample evidence of the mismanagement of the former Soviet-type economies.[2] But economists also recognise that efficiency is not the only objective that society has; the market system is efficient but it is also cruel; a planned system may be less efficient but more egalitarian.

That the operation of the labour-market for singers has its cruel side is indisputable. The risk of not earning a reasonable living, the uncertainty about employment, which can persist throughout a singer's life, and the gap between those who succeed and those who do not are the ugly side. But they also seem to be endemic in its operation. What is at the bottom of these problems is the oversupply of singers.

Oversupply and Market Failure

Whether or not there is an oversupply of singers is a very tricky question. There are three issues that must be taken into account and they are all interrelated; whether there is an oversupply of singers; whether the earnings of singers are lower than those of workers with comparable training and educational achievements; and whether the rate of return to training as a singer is lower than that to other types of higher education. These issues are interrelated because oversupply (or, in economic terms, excess supply at the ruling rate of payment) tends to depress earnings and leads to unemployment, and these tendencies reduce the rate of return to training. In a perfectly competitive labour-market in which all workers are motivated purely by financial incentives, the rate of

---

[2] For the way that this manifested itself in the opera and music world of the USSR see Vishnevskaya (1984).

payment would settle at an equilibrium rate at which supply would equate demand in all sectors of the economy and the supply of labour to each sector would allocate itself so that rates of return to different types of training and higher education would equalize. Then net earnings would be the same throughout the economy.

But despite a considerable degree of competition in the labour-market for singers, it is not perfectly competitive. There are considerable information problems and elements of risk that prevent the market from working perfectly. Moreover, the desire for psychic income on the part of singers—the pleasure gained from doing self-expressive work that one loves—means that they are not motivated only by financial incentives and this must be taken into account in relation to both the supply of singers and the comparison of their earnings with those of other workers. It is useful to summarize the findings on these topics separately and draw tentative conclusions about them in turn.

Comparison of Singers' Earnings with those of Comparable Workers

The main difficulty in making this comparison is that though there are official statistics on average earnings for non-manual workers, broken down by age and sex, in Britain there are no official statistics on singers' earnings and survey data are not reliable. (This is equally true of statistics on employment; while there are many signs of unemployment and underemployment in the singing profession, it is impossible to measure unemployment.) So, despite efforts to collect information on singers' earnings, it is not possible to state categorically whether, in general, singers earn as much as or less than other workers. It is clear that a few singers earn very much higher incomes than the average worker and also that many singers earn incomes below average earnings in the economy but it is not possible to be more precise. There is, however, evidence to suggest that singers' earnings have not risen at the same rate as the Average Earnings Index, suggesting that, whatever the initial position, singers have fallen behind other workers.

The issue is further complicated by the need to compare earnings by age and by sex. Again, without better data on singers' earnings it cannot be done. The median figure for men and for women in the cross-section data of singers' earnings presented in Chapter 5 compared almost exactly with average earnings of non-manual men and women in all occupations in the age group 25–9 in 1988; but it was supposed that the singers in the cross-section were older and so this comparison was not a fair one. However, the evidence certainly does suggest that women earn more in the singing profession than they would in other occupations and this may well explain why so many women seek to enter the singing profession.

The Rate of Return to Training in Singing

Here again a proper calculation is ruled out by the absence of proper earnings data. However, there are reliable figures for forgone earnings (the earnings of people entering the labour force with one A-level were selected as the most appropriate measure), as well as on the costs of training singers. Using notional figures for singers' earnings, based on the information collected on a sample of singers' earnings, and the Gowrie figures of the cost of a place in a music college (Gowrie 1990), private and social rates of return were calculated which suggested that the private rate is lower than that for male university graduates of other disciplines. The chief reason for that lies in the length of a singer's training, which extends well into his or her twenties. The social rate of return is low because the cost of a place in music colleges is relatively high.

But this does not mean that public money is being wasted in training singers, at least most of it is not. Although singers cost more to train than other graduates, their training has value other than in their becoming singers; training in singing is a general training and can lead to employment alongside other graduates in the general graduate labour-market. A graduate who has a music degree or other such degree that includes singing from a university, polytechnic or College of Higher Education has cost more or less the same to train as any other arts graduate; if he or she goes to work outside the singing profession, his or her earnings will be comparable to those of any other graduate. Students from music colleges cost more to train and so the extra cost of getting higher education in a music college is 'wasted'—but if they then earn more in another occupation than they would have done as singers they pay more taxes and so contribute more to society.

Is there an Oversupply of Singers?

The answer to this question has to be that it is not possible to say. If it could be shown that singers earned much lower earnings and had higher rates of unemployment than other comparable workers and hence had a lower rate of return to training, the answer would certainly be yes—but the concrete evidence is not there. But even if it could be proved that there is an excess supply of singers, what conclusions should one draw? On the one hand, it suggests that subsidy to training singers is too high and too many singers are being trained but on the other hand, it can be argued that not only is it equitable that singing students have the same opportunities to enter the profession of their choice as other graduates, but also that oversupply also benefits society by ensuring that it has the best possible choice of talented singers. So the issue has to be seen from two points of view, that of the singing profession and that of society at large. For the great complication in the discussion of whether or not there is an oversupply of singers is that, while there may be far more competent singers than the market can absorb, there is an apparent shortage of highly talented singers. It is therefore necessary to try to define the role of talent in the market for singers.

The Economics of Talent and Superstardom

Talent is a very difficult concept; it cannot be measured objectively and no one agrees on what it is fundamentally. However, it has various economic features—it takes up resources to search it out; it is highly rewarded financially and in other ways; and it may be a socially determined economic phenomenon. The search process for talent includes competitions, auditioning, and the training process itself. Information about singers and their talents is provided by singers' agents and managers, singers' brochures and audition tapes, and by performances at every stage of the singer's career from college days on. One of the economic functions of the higher education system is to provide information about students' abilities but in the case of the talents of singers it is not very effective as a screening device; paper qualifications do not play the same screening role in the market for singers as they do in other labour-markets. The costs of the search for talent and information about talented singers are therefore borne by employers in a range of performed arts organizations, by singers (through direct expenditures or via agents' fees) and to some extent by state subsidy to the training institutions. In some cases the costs of the search are higher than the perceived benefits, for example, fixers of choirs and choruses do not audition singers because they can find out on a one-off job how talented they are—if they are not suitable, there is little harm done. On the other hand, an opera company could hardly take that risk with a principal role and so must spend time and money in the search. The benefits of the search for talent have to be balanced with the costs.

The absence of an objective measure of talent opens up the possibility that talent is socially determined by the behaviour of employers or promoters on one side and consumers on the other side of the market. There are strong incentives in the market to identify and promote talented singers. Not only do employers have search and information costs, consumers have them too. The attempt to minimize these costs leads to both sides using price as an index of quality, i.e. the performance fee is treated as an indicator of how good the singer is—rising fees indicate rising stars and rising stars are more in demand. Although this seems contradictory in the sense that minimizing costs suggests opting for less expensive singers, it is because information costs are high to consumers, who have alternative uses for their time, that they are better off paying more to avoid disappointment. This causes the labour-market for singers to concentrate on a few singers. Once this process of concentration has started it can snowball in the individual case, even without there being a manifest talent to start the ball rolling. Convincing the public that you are good, according to this view, is what it takes; but any 'objective' information, such as being a competition-winner, will be latched on to by the market because of its value to employers/promoters and consumers. Fees are the signal about talent, real or fabricated, to the market; the fee structure in the market acts as a ranking and singers do not allow their position in the rank order hierarchy of talent to slip, even if they work less as a result.

In general, high fees go with considerable popularity and so with many performances; thus top singers earn very high incomes. At the very top, the superstars are enabled by developments in media technology to earn very high incomes indeed because they can reach mass markets not available to them in live performance; developments in microphone technology could eventually allow mass audiences for live performances by classically trained singers as well as for pop singers, but at present the technical constraints of live performance limits the size of audiences that singers can reach and the number of performances that they can do. This waters down the perceived tendency in some superstar professions for the market to concentrate on a very few superstars. These features of the market for singers arise on the demand side but they also affect the supply side. The chance of making very high earnings acts as an incentive to enter the singing profession to students who are prepared to take the risk. The theory of superstars also predicts that successive developments in media technology, with their increasing economies of scale, will widen the gap between superstars and the rest of the profession. Although the full effects may be slower in coming to concerts and opera than, say, to comedy or sport, there are signs, such as the 'Three Tenors Concert' and Nigel Kennedy on 'Top of the Pops', that they are fast approaching in the classical field.

The increasingly higher payments to superstars will act as an even greater lure to risk-taking student singers; as the prizes in the lottery of the singing business get higher, more will wish to enter and there will be greater pressure on places in institutions training singers. But the demands on singers' skills will also be greater, requiring training of a higher standard. The knowledge that fewer students will eventually succeed may cause music colleges to accept a lower student intake with a higher expenditure per student. Younger and better looking singers will probably be in greater demand, since in televised performances, at least, appearance in close-up can be more important than the sounds; also greater acting skills will be called for; these tendencies have already become evident in live performance.

Performing arts organizations will increasingly have to compete with media organizations for singers whose fees will rise fast once they get work in recording. One way round this is co-operation between TV, video and recording companies and the orchestras, opera companies, and other live promoters. This could, of course, be dangerous for singers because the organizations could conspire to depress singers' fees, as their combined power would be considerably greater than before. There is no convincing evidence so far that the growth of recording (visual and aural) has displaced the demand for live performances (it may even stimulate it) so there is no reason to suppose that the total demand for singers would fall, but those singers who only do live performances will probably earn even less in future in relation to those who do recorded performance—again the earnings gap will widen.

Earnings from recording will clearly form a greater proportion of total earnings than at present, though only for the chosen few. Where a live performance is recorded (aurally or visually) the singers get extra payments, thus increasing their earnings from a given effort. Earnings from recordings are spread more evenly over a lifetime than earnings from live performance because royalties and repeat fees continue over a period of years (the recording fee, of course, is paid at the time of the recording). This would reduce some of the uncertainty of singers about their ability to earn money over a lifetime. The conclusion to be drawn from this is that the growth of media technology—TV, video, recordings, use of stage microphones—will lead to greater concentration of the market on even fewer singers, increasing the earnings gap and making for even higher prizes for the successful few. But this has a feedback effect on the supply of singers.

One of the effects of the growth of media technology is the greater internationalization of both the production of music and opera and of consumers' tastes. It affects the market for singers in Britain in a number of ways: singers have to establish themselves on the international market to compete as stars; payments to singers therefore become geared to international supply and demand and so Britain has to compete with other countries for singers (including for British singers). Establishing yourself as a singer abroad means entering foreign competitions, travelling abroad for auditions, going abroad to work even when you would prefer to stay in Britain, and so on. It therefore raises the costs of searching for work and of information. British fees are already lower than those paid in many other European countries (though the apparently higher fees abroad may not be so much higher in real terms once differences in costs of living, taxes, and the like are taken into account). Undoubtedly the lower level of subsidy to the arts in Britain is one factor behind lower fees. However, there is another side to this, which is that lower fees should attract recording and other such work to Britain and make British singers more competitive abroad.

The Individual's Decision to Become a Singer

The above analysis suggests that the decision to become a singer is going to get more difficult; the top prizes may get higher but so will the cost of trying to win them. And not only will the chance of success be slimmer, the cost of failure will be higher. The greatest difficulty is that all the necessary information cannot be obtained—the most crucial piece of information, the individual's own chances of success, cannot be predicted and so uncertainty must always enter the equation. The more information students can get about the costs of and returns to training, the better they can make the decision. It is hoped that this book contributes to that. Unfortunately, however, students of singing do not seem to show much interest in these questions.

Many students (including those studying economics!) seem to take a sort of manpower-planning point of view; they seem to believe that there is a set training for a job—any job—and that if they get a place in an institution of higher education there must somehow be a job at the end of the course. Many of their teachers probably think that too. The greater the supply of places to study singing, the more such students will be encouraged to think of making a career in singing. If, as I believe will happen, the expansion of higher education leads to more and more places on courses in which singing is offered, this will encourage more and more students to dream of becoming singers. Everyone wants to be an artist and there is a huge pent-up demand for all types of artistic training courses. The least sign of hope—getting a place at college, a student grant, winning a competition—is interpreted as positive affirmation. This is not an argument for not subsidizing the training of singers, but it suggests that care is taken in doing so, so as not to offer incentives to people who stand little chance of success. The curious feature of the system of state subsidy to training singers is that it offers them most money at the stage at which least is known about them. For it is considerably easier to get a grant for an undergraduate course than it is for a postgraduate one—yet postgraduate singers must surely be a safer bet. Subsidy to singers needs to be much more flexible if it is to be cost-effective.

Society's Decision to Subsidize the Training of Singers

The other side of the coin is the benefit to society of having an oversupply of singers. Given the extent of the information problem about talent, it is surely the case that the greater the competition between singers, the more likely it is that talented singers will emerge. Oversupply leads to the maximization of the quality of singers from the audience's point of view and it is an efficient, if cruel, way of selecting talented singers. And though competition is harsh, it is probably more acceptable than a bureaucratic selection of artists would be.

In so far as the objective of subsidy to the arts is to ensure high quality work, support for 'too many' trainee artists is rational and efficient. If the objective of subsidy to higher education is equality of opportunity, then providing subsidized places in singing, music, and other arts, regardless of the outcome in terms of subsequent employment, is equitable. If students truly understood the implications of the decision they make when they begin to train as a singer, nothing more need be said. But do they?

What this study suggests is that there are two possible scenarios for the formal training of singers. There is one in which resources are concentrated in specialist music colleges (or other similar institutions), which train relatively few students with high potential in long and expensive courses. The other is a scenario in which many students have the opportunity to do less expensive undergraduate courses (degrees and diplomas) in Music, Creative Arts, Performance, and the like in universities, polytechnics, and Colleges of Higher Education which include singing, and then move on to specialist postgraduate courses for professional training. If all the students who trained as specialists from the beginning (the first scenario) make successful careers, their lifetime earnings would be sufficiently high to cover the costs, to themselves and to society, of the higher costs of specialist training. But failure would be very costly to both singer and society. On the other hand, the second scenario requires less assurance of individual success, and could still produce the same results in the long run, and probably with less risk. These two scenarios are not mutually exclusive and they co-existed during the period in which this study was undertaken, but there was no co-ordination of formal training. But formal training is only part of the way that singers learn their art; on-the-job training is also important and there was little financial support for that in 1990. The costs of that were borne by the singers with only limited help from subsidy and with help from private foundations.

Whatever the system of training singers, in the end the test of its success depends on the labour-market. That is why the question of training cannot be divorced from the question how the labour-market for singers works, and why it works as it does. The economic aspects of the singing profession cannot be avoided.

APPENDIX 1

Tables

TABLE 1. *The output of singers from music colleges, 1988*

Name of College	Total	Male	Female	Under-graduates	Post-graduates
Birmingham Conservatoire	9	3	6	—	—
Guildhall School of Music and Drama	44	22	22	19	25
Royal Academy of Music	34	12	22	10	24
Royal College of Music	30	8[a]	9[a]	17	10[b]
Royal Northern College of Music	30	14	16	—	—
Royal Scottish Academy of Music & Drama	30	3[a]	7[a]	10	20
Trinity College of Music	22	7	15	—	—
Welsh College of Music and Drama	8	2	6	—	—
TOTAL	207				

Notes
Not all colleges provided information about numbers leaving and some of the figures are intake figures for singing as first study. The elastic length and nature of courses in music colleges makes the notion of student output and the distinction between postgraduate and undergraduate qualifications difficult to draw. Although the length of undergraduate diploma and degree courses is fixed at three or four years, the length of time that students spend on them may vary with extension years being added on, repeat years and so on.

[a] These figures are for undergraduates.
[b] This figure is for the postgraduate opera course.

Source: Interviews with Heads of Singing and questionnaire to Music Colleges.

APPENDIX 1

TABLE 2. *Estimated output of graduates from courses in which singing formed the first or main study, 1988*

	No. with music dept.	%age with singers	Reported average no. of singers	Estimated total ouput of singers
Universities	34	50	3	51
Polytechnics	9	78	4	28
Colleges of Higher Education	20	80	4	64
TOTAL				143

Source: Postal questionnaire survey.

TABLE 3. *Distribution of hours worked in teaching singing by members of AOTOS, 1989*

No. of hours	%age of respondents
Under 10	18
10–20	34
21–30	26
31–40	16
Over 40	6

Source: Postal questionnaire survey.

TABLE 4. *Age-range of private singing pupils of members of AOTOS, 1989 (%ages)*

Sex	Age-range (expressed in %ages)			
	16–24	25–35	36–50	Over 50
Males	28	35	23	15
Females	34	31	25	10

Source: Postal questionnaire survey.

TABLE 5. *Overseas student fees in music colleges, 1990/1*

Name of college	Type of course	Fee (£)
Guildhall School of Music and Drama[a]	All music courses	4,272
Royal Academy of Music	Performer's	6,600
	Advanced	8,100
Royal College of Music	Opera Training	6,480
	Preliminary Opera	5,520
	Advanced Study	4,470
	GRSM	5,580
	B.Mus or M.Mus	5,850
Royal Northern College of Music		5,500
Royal Scottish Academy of Music and Drama[a]	Diploma/Degree and Advanced	4,200
Trinity College of Music		5,100

[a] Fees for 1989/90.

Source: College prospectuses.

TABLE 6. *Unit costs for the Royal Academy of Music, 1988/9*

Course	Unit Cost (£)	
	Using student hours as the basis of apportionment	Using teaching hours as the basis of apportionment
Advanced Singing	6,123	5,974
Advanced Opera	7,691	7,248
GRSM Singing	7,007	6,305
Performers Singing	6,134	5,708
Performers Opera	7,289	6,957

Source: Gowrie (1990).

TABLE 7. *Average gross weekly earnings (£s) full-time, non-manual, 1989*

Age-range	Earnings (£s)	
	Males	Females
18–20	138.5	125.3
21–4	211.5	165.8
25–9	276.7	206.2

Source: Department of Employment (1989: Part E, table 124).

TABLE 8. *Guide prices (£s) in higher education in 1990/1*

Subject area	Price (£s)	
	UFC	PCFC
Engineering and Technology, Physical Sciences	4,600	3,700
Education	3,500	3,057[a]
Creative Arts	3,300	3,500[a]
Humanities	2,800	2,303[a]
Business and Administrative Studies	2,800	2,494[a]

[a] The 1989/90 median price increased by the Treasury's forecast GDP deflator of 4%, not including the London Weighting Allowance.
Source: UFC and PCFC data.

TABLE 9. *Attendances (000s) at principal music venues in London 1984/5–1989/90*

Venue	Attendance (in 000s)					
	1984/5	1985/6	1986/7	1987/8	1988/9	1989/90
Barbican Centre	486	540	501	523	506	495
Purcell Room	60	66	64	64	58	64
Queen Elizabeth Hall	211	224	196	192	176	176
Royal Albert Hall (Proms)	206	212	207	218	214	—
Royal Festival Hall	716	704	680	687	672	660
Wigmore Hall	95	98	101	113	107	116
TOTAL	1,774	1,844	1,749	1,797	1,733	—

Source: Eckstein and Feist (1991).

TABLE 10. *Number of opera performances and seats sold 1984/5–1989/90*

Venue	%age increase in average ticket yield 1984/5–89/90	1984/5		1985/6		1986/7		1987/8		1988/9		1989/90	
		P	S	P	S	P	S	P	S	P	S	P	S
English National Opera	71	164	291	202	390	207	389	214	411	208	424	209	375
Glyndebourne Touring Opera	86	33	45	25	38	31	36	31	35	36	42	37	50
Kent Opera	69	42	28	43	36	38	32	39	36	42	28	37	23
Opera North	50	114	115	84	89	89	91	90	106	98	121	124	150
Royal Opera	105	133	262	132	248	120	227	144	270	128	245	147	288
Scottish Opera	—	90	94	94	99	116	130	107	119	122	157	101	129
Welsh National Opera	83	114	144	96	121	98	136	77	105	113	157	115	151
TOTAL		690	979	676	1,018	699	1,041	702	1,082	747	1,174	770	1,166

Notes:
P = Performances.
S = Seats Sold (in 000s).
Source: Feist and Hutchison (1990*a*).

TABLE 11. *'The choral society problem': trends in choral societies' outlays on singers and orchestras for the same work.*

	Soloists		Orchestra	
	Fees (£s)	Index	Fees (£s)	Index
Choral Society A:				
1985/6	4,600	100	5,200	100
1986/7	4,450	97	6,200	119
1987/8	5,550	121	6,600	127
1988/9	6,800	148	6,930	133
Choral Society B:				
1986	3,800	100		
1987	5,400	142	10,190	100
1989	6,500	171	10,900	107
Choral Society C:				
1979/80	430	100	700	100
1980/1	710	165	2,100	300
1981/2	750	174	1,270	181
1982/3	1,030	240	2,350	336
1983/4	700	163	2,560	366
1984/5	870	202	2,630	376
1985/6	1,430	333	2,880	411
1986/7	2,740	637	3,610	516
1987/8	1,280	298	3,460	494
1988/9	1,700	395	3,610	516
Choral Society D:				
1978/9	145	100	700	100
1979/80	715	493	720	103
1980/1	615	424	1,110	159
1981/2	1,040	717	950	136
1982/3	830	572	1,540	220
1983/4	930	641	1,550	221

Notes: Choral Society A used top singers and the same orchestra throughout; Choral Society B used the same singers, conductor, and orchestra throughout; Choral Societies C and D used different singers and orchestra in the different years for which figures are given.

Source: NFMS data, collected by the author.

TABLE 12. Cost indices for opera companies in Britain 1970/1–1980/1

	Weight	1970/1	1971/2	1972/3	1973/4	1974/5	1975/6	1976/7	1977/8	1978/9	1979/80	1980/1
Retail Price Index	(100)	100	109	117	129	152	190	219	249	270	313	364
Average Earnings Index		100	111	125	142	174	215	245	267	304	355	425
English National Opera:												
Aggregate Cost Index	(100)	100	111	118	130	151	198	221	245	274	318	365
Company principals and guest artists	7.1	100	114	114	156	154	237	241	238	244	311	366
Chorus, actors, and children	6.9	100	114	123	133	161	197	220	228	266	279	279
Orchestra	8.8	100	123	116	140	180	210	264	290	344	396	468
Royal Opera House:[a]												
Aggregate Cost Index	(100)	100	109	121	129	153	207	231	261	291	343	396
Labour (all except guest artists)[b]	(77.1)	100	109	120	129	152	202	223	251	282	336	392
Guest artists' fees	5.7	—	—	—	—	—	—	100	112	108	125	140
Chorus		100	111	114	125	131	176	211	231	284	233	399
Orchestra	12.6	100	108	119	126	163	195	210	263	312	359	424
Expenses other than labour	(100)	100	111	122	130	157	226	258	297	322	364	411

Scottish Opera:									
Retail Price Index		—	—	100	118	147	170	193	209	243	282
Average Earnings Index		—	—	100	123	151	173	188	214	250	299
Aggregate Cost index	(100)	—	—	100	122	147	166	186	215	251	295
Principal singers	3.5	—	—	100	124	147	187	224	239	293	413
Principals (regularly appearing artists)	9.1	—	—	100	124	147	187	209	252	281	344
Chorus	5.9	—	—	100	124	148	159	190	262	321	388
Chorus (SNO)	1.7	—	—	100	118	136	154	192	231	231	231
Orchestra	13.9	—	—	100	118	142	157	165	182	223	239
Welsh National Opera:										
Retail Price Index		—	—	—	—	100	115	131	142	165	192
Average Earnings Index		—	—	—	—	100	114	124	141	165	198
Aggregate Cost Index	(100)	—	—	—	—	100	119	132	143	168	204
Guest artists and conductors	5.8	—	—	—	—	100	114	100	114	143	229
Chorus, dancers, and actors	11.2	—	—	—	—	100	111	129	143	170	208
Orchestra	16.1	—	—	—	—	100	109	114	114	151	179

[a] The Royal Opera House figures include Ballet.
[b] Data on guest artists (in opera and ballet) were compiled from 1976/7 onwards. Therefore the labour index is based on 77.1 per cent of all other labour expenses.

Source: Peacock, Shoesmith, and Millner (1982: tables in ch. 7).

TABLE 13. *Cost-inflation indices for different art forms*

Year	Opera	Dance	Music	Drama	Retail Price Index
1970/1	100	100	—	—	100
1971/2	110	108	109	—	109
1972/3	119	115	116	—	117
1973/4	130	129	129	—	129
1974/5	153	156	152	—	152
1975/6	202	188	183	190	190
1976/7	227	220	207	221	219
1977/8	255	250	227	256	249
1978/9	284	287	255	290	270
1979/80	333	327	300	331	313
1980/1	388	388	345	395	364
Average annual inflation rate (%) 1975/6–80/1	13.75	15.25	13.5	15.5	13.5

Source: Peacock, Shoesmith, and Millner (1982: table 3.3).

TABLE 14. *Arts councils' subsidies (£ m) to music and opera in the UK, 1983/4–1988/9*

Arts council	Budget heading	Amount (£ m)						
		1983/4	1984/5	1985/6	1986/7	1987/8	1988/9	1989/90
Arts Council of Great Britain: England	Music	5.6	5.9	5.4	14.9[a]	15.7	—	—
	Opera	18.9	19.5	20.2	20.7	20.0	20.6	21.2
Scottish Arts Council	Music	1.8	1.8	2.2	2.4	2.5	2.6	2.7
	Opera	2.8	2.8	2.9	3.0	3.1	3.3	3.4
Welsh Arts Council	Music	0.5	0.5	0.6	0.6	0.6	0.88	0.87
	Opera	1.4	1.6	1.6	1.6	1.7	1.8	1.8
Arts Council of Northern Ireland	Music	0.6	0.6	0.7	0.7	0.8	—	—
	Opera	0.1	0.1	0.2	0.1	0.2	0.2	0.4
TOTAL	Music	8.5	9.0	8.9	18.6	19.5	23.3[b]	24.0[b]
	Opera	23.2	24.0	24.9	25.6	25.0	25.8	26.8

[a] The sudden increase is mainly due to the abolition of the Greater London Council causing the South Bank Centre to come under the auspices of the Arts Council of Great Britain.
[b] Great Britain (excluding Northern Ireland).

Source: Feist and Hutchison (1989: 7) and Arts Council 45th Annual Report and Accounts.

TABLE 15. *Median soloist's fees, paid by amateur choral societies in four UK regions, 1985/6–1988/9 (£s)*

Regional Arts Association	1985/6		1986/7		1987/8		1988/9	
	Amount (£)	n	Amount (£)	n	Amount (£)	n	Amount (£)	n
All	100–49		100–49		100–49		150–99	
Greater London	100–49	406	100–49	274	150–99	472	150–99	318
South-West	—	—	—	—	100–49	66	200–49	96
Yorkshire	100–49	211	100–49	202	100–49	233	100–49	229
Scotland	150–99	96	100–49	112	100–49	105	150–99	123

n = number of fee payments to soloists.

Source: Regional Arts Associations and NFMS data collected by the author.

TABLE 16. *Median soloist's fees paid by amateur choral societies in Scotland, 1978/9–1984/5 (£s)*

	1978/9	1979/80	1980/1	1981/2	1982/3	1983/4	1984/5
Amount (£s)	50–99	100–49	100–49	100–49	100–49	100–49	100–49
n	56	120	88	74	121	112	132

n = number of fee payments to soloists.

Source: NFMS data.

TABLE 17. Annual net earnings[a] of various singers (£s)

Singer	Earnings (£s)						No. of performances	Fee range (£s)	%age work in UK	%age work in Opera	Comments
	1983/4	1984/5	1985/6	1986/7	1987/8	1988/9					
1	—	—	—	4,000	10,000	18,000	60	250–600 (usually 500)	90	10	'Rising star' not quite ready for opera.
2	—	—	—	4,000	6,000	9,000	—	—	100	0	Not yet ready for opera.
3	—	—	—	—	11,000	13,500	—	—	—	70	
4	See comments			3,000	?	14,000	50	350–500	100	—	Earned approx. £10,000 in opera chorus in 1983/4–4/5.
5	—	—	—	3,000	?	11,000	35	300–500	—	some	
6	—	—	—	10,000	21,000	38,000	65	300–1,200 (opera 600–1,000)	66	—	
7	11,000	19,000	19,000	20,000	13,000	—	—	—	80	little	80% of work done in oratorio.
8	—	—	25,000	35,000	20,000	25,000	40	600	90+	80	Fee in 1985 £200–400, doubled in 3 years. In 1986/7 over 90 performances done at over £400.
9	—	—	19,000	21,000	31,000	58,000	40	700–4,000	50	0	Over 10% income from recordings. Usual UK fee £700, but got £4,000 abroad. Fees trebled over 4-year period.

10	—	—	78,000	74,000	127,000	60	3,000	50	50	UK fee rose from £200 to £3,000 over 4-year period.
11	27,000	26,000	59,000	70,000	—	—	—	50	75	Began to make breakthrough to international stardom.
12	—	—	27,000	35,000	32,000	35	700	80	80	Fee rose from £400 in 1985. Career tailing off.
13	9,000	26,000	13,000	15,000	—	40	400	100	90	These were gross earnings. 1984/5 income mostly abroad. Prefers UK work.
14	21,000	28,000	41,000	36,000	—	—	—	35	80	Career building very slowly. Has fulfilled career potential; career now steady.
15	60,000	96,000	66,000	147,000	167,000	—	—	50	75	Developing international star.

Notes: — indicates that information was not provided.
Average Earnings Index: 1985 = 100
　　　　　　　　　　　1986 = 108
　　　　　　　　　　　1987 = 116
　　　　　　　　　　　1988 = 126
　　　　　　　　　　　1989 (July) = 140

[a] Net earnings are pre-tax earnings net of travel and subsistence.

Source: Author's interviews and questionnaires, and data from agents.

TABLE 18. Cross-section of singers' earnings, 1988 (£s)

Singer	Gross income (£s)	Pre-tax net earnings (£s)	Expenses as a %age of gross income	Proportion of work in UK (%)	No. of performances	Average performance fee (£s)
A	100,000	60,000	40	60	75	900
B	75,000	45,000	40	—	100	750
C	32,000	—	—	10	76	400
D	25,000	13,000	28	60	20	65
E	21,000	13,000	38	90	452	—
F	19,000	8,000	58	95	195	85
G	16,000	10,000	38	80	—	—
H	15,000	13,000	13	70	108	70
I	14,000	10,000	29	90	20	210
J	12,000	10,000	17	75	60	100
K	12,000	6,000	50	98	424	—
L	10,000	9,000	10	100	12	—
M	8,000	—	—	100	—	—
N	9,000	5,000	44	90	70	35
O	9,000	6,000	33	60	150	20
P	8,000	5,000	63	100	15	55
Q	6,000	1,000	83	100	7	85
R	5,000	3,000	40	90	35	65
S	4,000	3,000	25	80	20	120
T	1,000	—	—	100	4	50

Source: Author's interviews and questionnaires, and data from agents.

TABLE 19. *One singer's earnings and tax-deductible expenses by age, 1977–1989*

Age	Gross income (£s)	Tax-deductible expenses (£s)	Net income (£s)	Other	Retail Price Index
23	2,000	1,100	900		100
24	1,400	?	—	Received 3 months' unemployment benefit	116
25	2,000	1,000	1,000		135
26	4,300	1,900	2,400		150
27	5,600	4,500	1,100		166
28	4,800	2,900	1,900		196
29	5,700	5,200	500		219
30	6,800	4,500	2,300		237
31	9,200	7,300	1,900	Expenses included tax withheld abroad	248
32	25,600	15,400	10,200	Taxes withheld abroad = £5,000	261
33	12,700	7,100	5,600		276
34	12,100	10,300	1,800		286
35	15,000	?	—		298

Source: Author's interview with the singer.

TABLE 20. *Average annual earnings of men with A-level qualifications by age-group, 1987 (£000s)*

No. of A-levels	Age-group					
	16–19	20–4	25–9	30–9	40–9	50–9
2+	3.4	7.1	11.1	11.3	22.1	14.3
1	3.7	7.7	9.2	11.2	13.5	8.2

Source: DES unpublished data.

TABLE 21. *Net-earnings differentials (£s): A notional singer compared with males with one A-level*

Age	Direct costs of formal training and tax-deductible expenses (singer)	Earnings forgone by singer (male earnings with 1 A-level, 1987)[a]	Gross earnings of a notional singer, 1987[b]	Net-earnings differential of notional singer: private rate calculation basis[c]	Net-earnings differential of notional singer: social rate calculation basis[d]
(Column 1)	(Column 2)	(Column 3)	(Column 4)	(Column 5)	(Column 6)
18	−6,000	−3,700	2,000	−1,700	−9,700
19	−6,000	−3,700	2,000	−1,700	−9,700
20	−6,000	−7,700	2,000	−5,700	−13,700
21	−6,000	−7,700	2,000	−5,700	−13,700
22	−1,000	−7,700	4,000	−4,700	−4,700
23	−1,000	−7,700	5,000	−3,700	−3,700
24	−1,000	−7,700	6,000	−2,700	−2,700
25–9	−1,000 × 5	−9,200 × 5	9,000 × 5	−1,200 × 5	−1,200 × 5
30–9	−2,000 × 10	−11,200 × 10	13,000 × 10	−200 × 10	−200 × 10
40–9	−2,000 × 10	−13,500 × 10	17,000 × 10	+1,500 × 10	+1,500 × 10
50–9	−2,000	−8,200 × 10	20,000 × 10	+9,800 × 10	+9,800 × 10

Notes: Minus sign denotes costs, plus sign denotes positive net amounts.

[a] See Table 20, above.
[b] Students are assumed to have received a £2,000 maintenance grant for all four years. The other figures are notional earnings of singers.
[c] Column 5 is calculated by Column adding Columns 2, 3 and 4. In the private rate of return calculation it is assumed that singers' fees are paid by the local authority.
[d] For the social rate of return student maintenance grants are ignored as they are a state transfer payment. The cost of the place is subsidized and therefore must be taken into account for the social rate of return.

TABLE 22. *What a singer would have to earn at an interest rate of 10 per cent to make training a worthwhile financial investment*

Age	Principal (£s)	Interest (£s)	Earnings forgone (£s)	Total (£s)[a]
22	17,919[b]	1,792	7,700	9,492
23	19,711	1,971	7,700	9,671
24	21,682	2,168	7,700	9,868
25	23,850	2,385	9,200	11,585
26	26,235	2,624	9,200	11,824
27	28,859	2,886	9,200	12,086
28	31,745	3,175	9,200	12,375
29	34,920	3,492	9,200	12,692
30	38,412	3,841	11,200	15,041
31	42,253	4,225	11,200	15,425
32	46,478	4,648	11,200	15,848
33	51,126	5,113	11,200	16,313
34	56,239	5,624	11,200	16,824
35	61,863	6,186	11,200	17,386

[a] 'Break-even' Earnings = Interest + Earnings Forgone.
[b] The lump sum needed to compensate a singer for training rather than going to work is £17,919. This was calculated as follows, using the data in Table 21, above:

Year 1 Net earnings forgone £1,700 compounded for 4 years at 10% = £2,489

Year 2 Net earnings forgone £1,700 compounded for 3 years at 10% = £2,263

Year 3 Net earnings forgone £5,700 compounded for 2 years at 10% = £6,897

Year 4 Net earnings forgone £5,700 compounded for 1 year at 10% = £6,270

TOTAL   = £17,919

APPENDIX 2

Calculation of the Rate of Return to Training in Singing

THE PRESENT VALUE FORMULA AND THE INTERNAL RATE OF RETURN

Using the example of the bank deposit of £100 with a rate of interest of 10 per cent, it is possible to view future interest payments from the time of the deposit, i.e. the present. A deposit of £100 will earn £10 in one year at an interest rate of 10 per cent. But one could also ask how much would need to be deposited at an interest rate of 10 per cent to earn £10. The answer is obviously 100 and that is called the present value, V_1, the subscript denoting the length of the investment.

$$V_1 (1 + 0.10) = £110.$$

Therefore

$$V_1 = £110/1.1 = £100$$

is the present value. After two years the deposit will earn £21 and the total value is £121; then

$$V_2 = £121/1.1^2 = £100$$

is the present value. In general, if earnings, E, last for t years, say from 18 to 60 years of age, we may calculate the present value, V, as the total of the future stream of earnings over a period of forty-two years at an interest rate, i.

$$V = \sum_{t=18}^{60} E_t/(1 + i)^t$$

is the present value formula. (Σ means 'the sum of').

In the above example, it was assumed that the interest rate was 10 per cent and the question was what V would be needed. But the problem of finding the internal rate of return is somewhat different. The question is this: given that it costs a certain amount, C, over a certain number of years t, to train a singer who will then earn a total E until retirement at 60, what rate of return will equalize the stream of future earnings and the stream of costs viewed in the present when a decision has to be made? (To reiterate, the decision is whether the investment in training is financially worth while.)

$$V = \sum_{t=18}^{60} E_t/(1+r)^t - \sum_{t=18}^{60} C_t/(1+r)^t$$

is the present value formula but now it is r, the rate of return, that we want to calculate and this can be done by setting $V = 0$.

So this gives us the rate of return formula that we need:

$$V = \sum_{t=18}^{60} (E_t - C_t)/(1+r)^t = 0$$

V can be solved from cost and earnings data over t years. If a singer began training at 18 and worked to 60, t runs from 18 to 60, i.e. $t = 42$ years; if the singer's training lasts four years, the first four years, $t = 1 - 4$, of the expression will be negative and only become positive when earnings exceed costs.

A NOTIONAL RATE OF RETURN TO TRAINING IN SINGING

In order to calculate the rate of return to training in singing, data are needed on singers' net earnings, on the costs of training, on the average training period and on the average number of years over which singers work. While we have good information on the costs of training and on the average period of formal study, there are no reliable data on lifetime earnings nor on the typical length of singers' careers. It is therefore impossible to make a correct calculation of the private or social rate of return to training in singing. However, what can be done is to suggest what it might be using fairly realistic notional earnings, assuming retirement at the age of 60; in addition, using the information on costs and the usual length of training, it is possible to indicate what singers would have to earn to make training financially worth while. Notice that in this calculation, very favourable assumptions are made about the singer's ability to get work immediately on graduation and thereafter.

Besides the data relating to singers, data on the lifetime earnings of 18-year-olds who have not undertaken higher education are also needed to estimate forgone earnings. Such cross-section data have been calculated for the DES from the General Household Survey for male workers during the 1980s, ending in 1987. The figures are given in Appendix 1, Table 20. In 1987 annual average earnings of men aged 16–19 with one A-level were £3,700, rising to £13,500 for men aged 40–9. In Table 20 men with one A-level earned slightly more up to the age of 25 (probably because they entered the labour force a little earlier), but those with two or more A-levels soon overtook them and earned more in later life. For purposes of comparison with singers we could take either age-earnings streams. Some singers have two or more A-levels and some one or no A-levels at all, singers on a

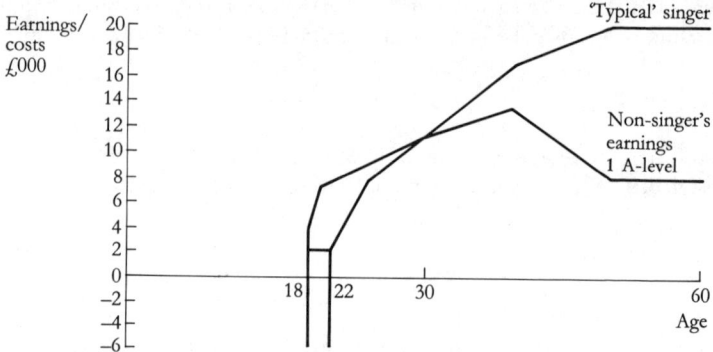

Fig. A2.1. Net Earnings Differentials of Singers

graduate course needing A-levels while those doing a Performers' Diploma do not necessarily need any. (The singers I interviewed nearly all had three A-levels but these singers were probably not representative of the profession as a whole.) It seems sensible, therefore, to take the earnings of men with one A-level as the group with which to compare singers. These earnings streams are the basis from which the net earnings differential and the present value of forgone earnings are to be calculated using the formula described above. This is illustrated in Figure A2.1.

For the notional calculation, it is assumed that singers enter at 18 and spend four years in a music college in which the cost of a place is £6,000 a year. The figure of £6,000 is used rather than the £6,400 published in the Gowrie Report (Gowrie 1990) for music colleges (see Chapter 2) because that figure was for 1989 and the earnings data are for 1987; in a rough calculation such as this, round numbers are all that are needed. All the data for the rate-of-return calculations are set out in Appendix 1, Table 21 and illustrated in Figure A2.1. Notice that the notional singer's gross earnings overtake the average earnings of a man with one A-level from the age of 30 onwards, when the singer earns £13,000 but net earnings delay the age earnings profiles from crossing until the age of 40 when the notional singer's net earnings are £15,000.

In Table 21, the cost of the place in a music college is entered in column 2 for the four years of formal training. Also in column 2 are notional tax deductible expenses of singers, assumed to be £1,000 from the start of work at 22 to the age of 30 and £2,000 thereafter. Column 3 is taken from Table 20 and represents the earnings forgone by singers, the opportunity cost or indirect costs of not entering the labour force at 18 with one A-level; column 4 gives notional earnings of singers and assumes that the student received a maintenance grant of £2,000. Singers are assumed to have gross earnings of £4,000 at age 22, rising to £20,000 a year by the age of 50; this latter figure is based on what was found to be the average earnings of singing-teachers in 1988. It should be stressed that these figures for gross earnings in column 4 are fictional but they are consistent with those reported in Chapter 5; if anything they are on the high side for 1987 and they are assumed to be regular, which is probably unrealistic in view of those survey results. The net earnings differential in column 5 is calculated by deducting column 3 from column 4 minus tax-deductible expenses in column 2 from age 22 on, i.e. when the notional singer starts work. In order to calculate the private rate of return the direct costs of formal training are ignored because they are subsidized and therefore do not form a cost to the individual, but they appear in the calculation for the social rate of return based on column 6 because they are a cost to society, being paid out of general taxation.

The figures in column 5 are the basis for the calculation of the private rate of return and those in column 6 are used to calculate the social rate of return using the formula given previously. The private rate of return is calculated from Formula 1 and the social rate of return from Formula 2 by solving for r.

Formula 1

$(-1,700)/(1 + r)^1 + (-1,700)/(1 + r)^2 + (-5,700)/(1 + r)^3 + (-5,700)/(1 + r)^4 + (-4,700)/(1 + r)^5 + (-3,700)/(1+r)^6 + (-2,700)/(1+r)^7 + (-6,000)/(1+r)^{8-12} + 15,000/(1 + r)^{23-32} + 98,000/(1 + r)^{33-42}$.

Formula 2

$(-9,700)/(1 + r)^1 + (-9,700)/(1 + r)^2 + (-13,700)/(1 + r)^3 + (-13,700)/(1 + r)^4 + (-4,700)/(1 + r)^5 + (-3,700)/(1 + r)^6 + (-2,700)/(1 + r)^7 + (-6,000)/(1 + r)^{8-12} + (-2,000)/(1 + r)^{13-22} + 19,000/(1 + r)^{23-32} + 98,000/(1 + r)^{33-42}$.

The private rate of return calculated from Formula 1 is 4 per cent and the social rate of return calculated from Formula 2 is 1.7 per cent.

WHAT A SINGER WOULD HAVE TO EARN AT AN INTEREST RATE OF 10 PER CENT IN ORDER TO MAKE TRAINING A WORTHWHILE FINANCIAL INVESTMENT

The current rate of interest can be used as a discount rate with the present value formula to calculate what a singer would have to earn to make training in singing a worthwhile financial investment. Imagine the following scenario: a rich uncle offers to give you a lump sum which could either be put into a savings account earning 10 per cent or spent on financing your training for four years in a music college. He agrees to compensate you for forgone earnings for the four years while you study but not thereafter. How much would you have to earn to break even as a singer compared to what you would earn with one A-level if you went to work at 18? Looking at Table 21 column 3, you can see that at age 18 you could earn £3,700 by going out to work, less £2,000 for the student maintenance grant you get as income, i.e. the net figure you need to compensate you for going to college is £1,700. That is what your uncle has to give you in Year 1.

If, instead of going to college, you were to open a building society account paying 10 per cent interest, by the end of the first year you would have an interest payment of £170 as well as the principal sum of £1,700, in all £1,870. Similarly after the second year you would have £187 interest from the Year 1 principal sum and in addition he needs to give you a further £1,700 because you are not working; this could also earn interest. In Year 3 at age 20 you could earn £7,700 by working, so taking the student maintenance grant into account, you need £5,700 to compensate you for not working; again the same thing in Year 4. All this time, if you had gone to work and kept the money in the building society account instead, you would be earning interest on the annual sums of money from your uncle. In fact, the lump sum needed to cover your indirect costs of earnings forgone would total £17,919. (This is worked out in detail in Table 22.) This is the amount that you 'spend' going to college to study singing, your part of the investment in your human capital. (The government also spends £6,000 every year you are there too, but you need not take that into account.)

If you had not gone to college but had put the whole amount into the building society and left it there, it would go on earning interest at 10 per cent. At the age of 22 you would now be earning £7,700 by working plus the interest on your building society deposit, which is 10 per cent of £17,919, i.e. £1,791.9 (or £1,792 in round numbers). So in the first year that you leave college, you must earn £7,700 + £1,792 = £9,492 to do as well as you would have done by going out to work at age 18. The remaining calculations up to the age of 35 are shown in Table 22. By the age of 35, you would need to earn £17,386 to break even. This imaginary scenario is presented here to demonstrate in detail what is involved in the notion of investing in human capital by training rather than in monetary financial investments. It should make clear the exact financial sacrifice that is made by anyone undertaking training or higher education.

APPENDIX 3

The Training and Employment of Singers in Germany[1]

Many British singers work in Germany and it is often held up as an example of a country in which art and artists are taken more seriously than they are in Britain. I therefore visited West Germany in 1990, before reunification, as part of the research for this book.[2] The Federal Republic of Germany probably has the biggest concentration of singers in the world. In 1987/8 it had ninety-three theatres in fifty-three different cities in which opera was performed. Some of the opera-houses are owned and administered by the individual states of the Federation while others are owned and administered by the municipal authorities. Similarly, music schools are provided by many states and municipalities. This enormous collection of opera-houses and music schools reflects the historical development of the FDR as a union of separate states. The reunification of Germany has by now, of course added considerably to the total of these organizations. However, this appendix reflects the situation in former West Germany in 1990.

Because cultural and educational provision is decentralised, it is usually very difficult to make any generalizations about the arts in Germany. However, in the case of opera and spoken theatre it is not a completely impossible task to get a picture of the training and employment situation, as all theatres belong to the Deutscher Bühnenverein (German Stage Association), which collects annual statistics on all aspects of German music and spoken theatre for the whole of the FDR. This is an employers' association. In addition, the GBDA—the main trades union for theatre employees—produces a comprehensive yearbook (the *Bühnen Jahrbuch*) which provides information on employment in theatres, radio choirs, etc. and on training. Though by no means all singers are members of the GBDA, binding agreements are made between the GBDA and the Deutscher Bühnenverein on pay and working conditions to which employers adhere *de facto*, though not *de jure*. Thus it is possible to describe the employment situation, complicated as it is by many different factors, with some confidence.

---

[1] In this appendix, 'Germany' refers to 'West Germany' before reunification.
[2] Research on this topic was made possible by a grant from the Esmee Fairbairn Charitable Trust.

APPENDIX 3

THE TRAINING OF SINGERS

Although comprehensive statistics are available on employment, that is unfortunately not the case with training establishments. There are over eighteen Hochschule für Musik (music high schools, literally translated) in the FDR, listed in the *Bühnen Jahrbuch*, which perform the same function as music colleges and conservatories in Britain. In addition, there are many municipal music schools which also offer training for singers at a lower level and there are private teachers and academies besides. The high schools form part of the general provision for higher education and award a diploma which is equivalent to a university degree. The diploma covers solo singing for opera, concerts, and church music and is also a teaching qualification. Teachers of school music are usually taught elsewhere. In the state of Nordrhein-Westfalen, where research on this topic was undertaken, the music high schools were all unified until recently and some are still linked to each other. These schools all have the same requirements for the diploma, which are twelve semesters of study, there being two semesters in the academic year (one of fifteen weeks and the other ten) with set examinations and a laid down curriculum for each type of study. In Cologne, the biggest of these (and one of the biggest musical conservatories in the world with a total of 1,800 students) basic studies for singers comprise piano, harmony and counterpoint, etc., choir and Italian language and a one- to one-and-a-half-hour individual singing lesson a week, which continue throughout the six years. In the third semester, opera studies begin with the opera department. Examination requirements for opera are five completely studied roles (one modern), speech (dialogue), and one tragic role in a fully produced publicly performed opera; the examination requirements for concert singing are a concert of ten arias and ten Lieder. For the singing-teacher's diploma, students must have given sixty hours of individual and ensemble lessons as teaching practice.

Students usually enter at 18 with a pass in the Abitur (secondary school leaving diploma), though they can be accepted up to the age of 27 without Abitur, in the case of someone with exceptional vocal potential. Tuition is free (including for foreigners) and students may get maintenance grants. Students are usually 24 or 25 years old when they leave. The state of Nordrhein-Westfalen pays the teachers' salaries in the music high schools and they have contracts for either eighteen or twenty-one hours a week for the whole year, with a generous pension and health plan. (There are, in addition, some part-time teachers). The short number of weeks in the two semesters of the academic year leaves free time for performing, which many teachers do. In the Cologne music school in 1990 there were 210 singing students and 20 teachers, not all full time, with additional coaches, opera producers, and the like.

This was the situation in one state and I do not have the information from which to generalize from the position in Nordrhein-Westfalen to the other states and municipalities, the information is presented as merely interesting in its own right. Nordrhein-Westfalen has more opera-houses than any other single state in the FDR and the Cologne music high school is reputed to be the second biggest in Europe (Moscow being the biggest), so it is clearly not representative of other states.

OPERA STUDIOS

There are five opera studios attached to the opera-houses in Berlin, Cologne, Düsseldorf, Hamburg, and Munich. In them, students receive a mixture of formal and on-the-job training by doing small parts in the main houses and studying in the studio. They have contracts and were paid around 2,000 DM a month (about £8,500 a year) in 1990, of which half is a stipend for living expenses and hence tax free (which makes their pay equivalent to the minimum salary of 2,500 DM explained below). In addition, they are paid a small performance fee, 300 DM a month (about £100) in 1990 when they perform. Students typically spend one to two years in the studio, doing fifty performances a year of about six roles. They get unlimited coaching to learn roles, attending rehearsals of the operas in the main house and may get the occasional opportunity to step in if a singer is unable to do a performance. Some opera studios offer singing lessons as well.

The benefits of an opera studio are that young singers become involved in the day-to-day running of a large opera-house and so acquire professional standards of behaviour and learn from work with experienced singers. There is a danger of students being used as cheap labour, because after all, being part of the main opera-house means the studio serves its needs rather than those of the individual student, but the counter-argument is that they would otherwise be working under considerably less favourable conditions in a smaller opera-house. There is no in-between establishment, like the National Opera Studio, in Germany; nor is it likely that young singers would be able to freelance at the beginning of their career and so control the number of performances that they do. Thus the advantage of the opera studios is that they can nurture and protect young talented singers who need grooming for a solo career. There were about fifty singers all told in these studios in 1990, quite a few being foreigners.

EMPLOYMENT IN OPERA-HOUSES

All opera-houses in Germany have regularly employed singers both in the chorus and with the ensemble of soloists. At one time practically all major roles were cast from the ensemble (or company, in British terms) with occasional guest performers. Increasingly, however, many opera-houses, usually the bigger ones, have come to use guests and have accordingly cut down on the number of ensemble singers. This has come about for a variety of reasons; successful singers prefer to have free time for guesting in other opera-houses, because they earn more money that way, and audiences like the variety that guests provide and also the opportunity to hear singers with an international reputation—the effect of recordings. There are both economic and artistic reasons for this change; however, opera managements prefer not to hire guests because it makes planning and rehearsal schedules more difficult; they acknowledge, though, that having an ensemble large enough for casting all roles is a more expensive way of doing things. Thus this change has not been brought about by financial pressure on the part of managements but rather reflects the demands of singers for freedom to plan their careers. One artistic side-effect of the occasional use of guests is that they get practically no rehearsal for a production that is already running or in the repertoire—for a graphic description see Legge (1988: 46). To understand how this situation comes about, it is necessary to explain some features of the organization of opera performance (which apply, of course, elsewhere than in Germany).

The Stagione System

The stagione (Italian for season) system is that in which an opera runs for a set season of so many performances a week, with reasonable time between them for singers to rest. In its purest form, that would mean that one opera is rehearsed for several weeks, put on for its run, and then finished with. Then another opera is prepared and subsequently performed. Thus one cast of singers works for the season or run of the opera and then departs. This always has been the Italian system; in the old days the operas would almost always be new ones and thus there was no repertory (Rosselli 1984). In the stagione system the singers normally are all guests (freelance).

The Repertory System

In the repertory system different operas are performed throughout the year, or over several years, interchanging with each other. After the initial rehearsal period, the operas remain in the repertoire and are revived regularly; in Germany this often happens without further rehearsal. After several months or even years there would be only a limited rehearsal period (maybe one session) as singers (usually the same ones) would be expected to remember their parts and moves. This system requires a regular ensemble of singers who have been in the house over a long period of time and who work in several of the operas in the repertory, playing one role one evening and another another (sometimes the next evening). If a singer is ill and another has to replace him or her, the replacement, who is hired because he/she knows the role—there are no covers (understudies) in Germany, mainly because there are so many singers around—gets shown the set and told the main moves and goes on cold. The repertory system is the traditional German system; audiences are accustomed to seeing the same singers in different roles, sometimes growing old together, as singers might stay in one opera-house for the whole of their working lives, while audiences regularly attend for the whole of their lives!

The Semi-Stagione System

This is a mixture of repertory and stagione productions. Several new productions are given each year and then join the repertoire for future years while others drop out. This system requires a combination of guest and ensemble singers, the former probably doing the main roles with the remainder of the roles being filled by the company singers. It calls for some rehearsal time for revivals, as new guests are introduced to the production—often two to three weeks. The semi-stagione system is nowadays operated in most of the bigger German opera-houses, as well as in Britain and elsewhere.

These systems of organizing opera performance have implications for the demand for singers and lead to different types of contracts. Although this description refers to opera, it should be said that included in it are operettas, which all German opera-houses mount throughout the year. It does not include musicals, however, which in Germany are performed by actors.

SINGERS' CONTRACTS

All German opera-houses (with the exception of Bayreuth) are either state or city enterprises and accordingly singers, as well as all the other employees of opera-houses (musicians, dancers, actors, producers, administrators, etc.), are paid by the state or city authorities; in that sense they are civil servants or local government employees. Employment contracts are accordingly official documents and subject to bureaucratic regulations. Agreements between the Deutscher Bühnenverein and the GDBA have laid down standard forms for these contracts and there are essentially five types which relate to singers.

Chorus Contract

The chorus contract allows for a trial period, usually of one year, after which singers get a permanent contract. After fourteen years they obtain tenure, the right to be paid by the state or city authorities until 65, the retirement age. Within the fourteen-year period the contract can be terminated but in practice few are, at least on artistic grounds—a doctor's certificate is required to show that a singer is no longer able to sing properly!; reauditioning seems not to take place. After the fourteen-year period, an unsuitable singer has to be given continued employment but this need not be in singing—they could work in administration or on the stage door! Incidentally, this applies to dancers, actors, and orchestral musicians as well.

Chorus Contract with Small Parts

In this type of contract the singer will be required to undertake a certain number of performances of small parts within a stated period and within their Fach (see below), for which they are paid (perhaps £100 a performance in 1990). What 'small parts' are depends to some extent on the size of the opera-house and on the number of full-time soloists they have to deploy. In big houses with many contracted singers, small parts may be done by experienced soloists. This contract is used by medium and smaller opera-house managements to attract good young singers (the smaller the house, the smaller the number in the chorus and hence the more noticeable the contribution of the individual singer).

There are a host of rules and regulations surrounding the deployment and pay of choristers. All singers are entitled to free afternoons from 1 to 5 p.m. (unless there is a matinee performance, in which case this must be compensated for). In addition, they are paid extra for moving scenery, carrying heavy objects, make-up, certain types of dancing, speaking lines, etc. Choristers are entitled to one free day a week and they normally perform on Sundays. Of course, not all choristers are involved in every performance; some operas call only for a small chorus or a male chorus or, indeed, have none. Nevertheless, in a large opera-house which could have 300+ performances a year, they will be kept very busy; equally, those in small choruses are kept busy because all of them would have to work in a small house where in a larger house only a proportion of the whole chorus would be needed. The chorus and other regularly contracted singers and employees in opera get a forty-five-day annual holiday with pay and this is sacrosanct—if a singer is ill in the holidays they are entitled to working time off in lieu.

Solo Singers' Contracts

There are basically three types of contracts for solo singers, the normal contract (*Normalvertrag*) lasting two years usually, but one is also possible. Then there are the *Abendvertrag* (literally the evening contract) and the *Gastvertrag* (guest contract) in which singers are paid per performance or for a set number of performances of a particular role, and the part-time contract (*Teilspielzeitsvertrag*), which is essentially a mixture of the preceding types.

Normal or Fest Contract

This is often called a *Fest* contract (*Fest* meaning firm) and is the usual type of contract for ensemble or company solo singers in the repertory system. However, that being said, there are nearly as many variations within this type as there are evenings and afternoons in the week, since specified in the individual's contract is the upper limit of the performances within the correct Fach (literally 'department, pigeonhole') that that singer can be required to perform. The Fach essentially represents the type of voice and type of role for which it is suitable; for example, Susanna in *The Marriage of Figaro* is a lyric soprano role and so must be performed by a lyric soprano; it is one of the roles that is specified in the lyric soprano Fach, along with a list of others (see Legge 1988 for accessible details of this question; the definitive work on the subject is Kloiber's *Handbuch der Oper*). Thus a lyric soprano gets a job as a lyric soprano and a contract that specifies she is a lyric soprano, which means she can only be required to sing lyric soprano roles. Each opera-house has a certain number of posts for lyric sopranos and, of course, for singers in each other Fach. Singers can work out of their Fach if this is mutually agreed between them and the management and it is possible to change Fach if the voice changes (e.g. from lyric to young dramatic) or if the voice is no longer what it was; for example, an older singer who had been a dramatic or heroic singer may no longer be able to sustain a major role and therefore can change to a character Fach. What role is in which Fach is laid down by the pay agreement (*Tariffausschluss*) of the Deutscher Bühnenverein.

There are artistic aspects to the Fach system that are not part of this study. What is relevant here is that the Fach system has implications for the number of solo singers that are employed and what they are paid. Because some types of voices are rarer than others—dramatic tenors and basses being shortest in supply—favourable contractual arrangements are made to attract them, as well as higher salaries. A normal contract which specifies only thirty performances a year gives a considerable scope for guest appearances with other opera-houses (on a *Gast-* or *Abendvertrag*); the singer benefits by earning performance fees elsewhere and the management benefits by having the singer at its disposal when it wants him or her; permission has to be sought for specific dates of absence. The opera-house can offer its public good singers and gains in reputation from their successes elsewhere. In addition it does not have to pay singers as much for a thirty-performance limit contract as it does for a limit of a higher number of performances; this would make financial sense when the type of operas being performed do not often call for a particular Fach; that depends upon the balance of heavier to lighter repertoire being performed—a house doing a lot of operetta will not need many dramatic basses but it would need them for Wagner; it needs soubrettes in operetta but not for Wagner, and so on. At the other end of the spectrum, there are soloists who are engaged on unlimited performance contracts where singers, often young or unsuccessful older ones, can be required to perform every role within their Fach at every performance throughout the year, maybe a 100 or so performances a year. This is particularly likely to be the case in smaller opera-houses, though as these smaller city theatres do a mixture of opera/operetta, spoken theatre, dance, and musicals, this limits the number of performances that a singer actually has to do. In between the two ends of this spectrum, with unlimited performance contracts at one end and favourable limited contracts at the other, are a variety of normal contracts with performance limits. Here the generalization ends; however, examples are interesting. One big opera-house, which was mainly on the repertory system, had sixty people on *Normalverträge* and thirty-eight on *Abendverträge* in 1989/90; half the sixty had limits on their contracts, ranging from ten to seventy performances. Another major house on a semi-stagione system had twenty-three singers with a *Fest* contract and had hired sixty-five singers on *Abendverträge*. When singers on a limited performance normal contract do more than their contracted number of performances, they get paid extra on a pro rata basis (see below).

A normal contract is usually issued for two years in the first instance, though it can be for one year only and this would appeal to young beginning singers. Contracts can be terminated if a singer wishes to move on; singers can also be sacked within a fourteen-year period but, unlike the case with the choir contract, this is not so unusual; singers who are no longer suitable for any solo Fach can be moved on to an *Abendvertrag*, paid the same for a certain number of performances as they would have had on a normal contract and then dropped. If a singer is sacked outright they have the right to two years' unemployment compensation, as with all German employees. After fourteen years they, like choristers, acquire tenure and the same regulations apply to soloists as to choristers.

Abendvertrag

The *Abendvertrag* requires little explanation; the singer is paid for one evening or the number of performances of a particular role, either in their Fach or outside it. But, like freelance singers in Britain, they have no job security and are responsible for their own health, unemployment, old age and accident insurance. Under a normal contract, the opera-house pays half of that. As the employee's contribution to these policies is 25 per cent of their salary and the employer's contribution is also 25 per cent this provides a strong financial incentive to singers to have one foot in each camp, i.e. a normal (*Fest*) contract with plenty of free time for guesting under *Abendverträge*.

The part-time contract is a half-way house between being completely independent and having a full-time normal contract. According to German legal and accounting conventions, if a singer does more than six performances with one opera-house or spends more than fifty days a year there, they have a part-time contract. This entitles a singer to pro rata payments for health, etc. insurance as under the normal contract.

This completes the somewhat lengthy account of the main types of contracts. It is not exhaustive—only exhausting!

THE AMOUNT OF WORK IN OPERA

Because the labour-market for singers in opera in Germany is dominated by contracted employment, it is possible to give a figure for the number of jobs for singers in opera. Such a figure is provided in the *Theaterstatistik 1987/8* of the Deutscher Bühnenverein. The figure for choristers is straightforward; however the figure for solo singers is for those with a *Fest* contract only and therefore does not represent a count of the amount of work available for soloists defined, for example, as the number of singer-performances in the year, since many performances are done by guests. If we were to take the list of solo singers in every opera-house and count their number, this would double-count the same singers many or, anyway several, times over, because of singers on *Abendverträge*. On the other hand, if we could assume that every singer performing anywhere had a *Fest* contract somewhere, the count of singers on those contracts would identify the total number of singers employed in Germany. In fact, the true figure will be higher than that as there are freelance independent German and foreign singers working as guests and by the evening.

In 1987/8 there were 2,040 choristers and 1,174 soloists involved in 5,250 performances of opera and 1,909 performances of operetta—attended by audiences of 4.5 million for opera and 1.3 million for operetta. A quarter of all these took place in North Rhine–Westphalia.

The *Theaterstatistik* also provide information on each opera-house and it is therefore possible to compare the statistics with information provided in interviews with administrators. These show, as expected, that the chorus figure is a reliable guide to the number of jobs available but the figure for soloists underestimates the number of singers who have performed with the company. The extent to which it is underestimated depends on the opera-house's policy; in an opera-house which has a policy of using guests, the gap can be considerable. In Cologne, in the 1989/90 season there were twenty-three singers with *Fest* contracts and sixty-five more listed in the season guide, a ratio of one to three roughly; however, apart from enquiring about each of them, there is no way of knowing if a singer has performed for just one night or several. Another example is the Deutscher Oper-am-Rhein in Düsseldorf with over 100 soloists listed in the season guide of whom 65 have *Fest* contracts, thus the ratio of permanent jobs to the number of singers required for performances is much lower at a rough ratio of one to one-half. This difference means that it is simply not possible to make any reliable estimates between the balance of permanent jobs to freelance work.

The *Deutsches Bühnen Jahrbuch* for 1988 gives a count of soloists

238   APPENDIX 3

including those with part-time, guest, and *Stück* (for the piece, i.e. for a role in the run of performances of one opera) contracts; for the 1986/7 season this was 1,764. This probably comes nearer to the (unknown) true figure but, again, does not include *Abendverträge*.

PAY AND EARNINGS OF SINGERS

For all singers, soloists, and choristers, who are on *Fest* contract, there is an agreed minimum monthly pay; in 1990 this was 2,500 DM (just under £900); however, all choristers and some beginning singers are paid for thirteen months a year (a sort of built-in Christmas bonus) so that the minimum annual pay is higher—around £11,600 in 1990. For choristers, there are additional payments as outlined earlier which probably increase earnings by 5 per cent so that a chorister's minimum annual earnings are more or less £12,000. This was the minimum in 1990; in fact, the average pay of choristers is usually higher but the minimum is probably what a beginner would normally earn; deducted from this is 25 per cent for obligatory social security payments and also, of course income tax (and tax rates are higher in Germany than in Britain).

The pay of choristers varies from house to house and also depends on the singer's age and vocal quality—it is quite possible for salaries to fall with age. Information provided in the *Theaterstatistik* makes it possible to work out average earnings of choristers for each individual house as well as for the whole of West Germany for 1987/8. However, this overestimates the true picture to some extent, which should be borne in mind. The average annual earnings for choristers were just over £20,000 with the big houses paying above average (as one would expect)—Berlin, the biggest, paying out about £29,000 on average to its 110 choristers and the smaller houses paying close to the minimum, with some well below the minimum if the choir were part time or amateur. It should be said in parentheses that even the bigger houses use amateur extra choristers upon occasion, there being no equivalent in Germany of British freelance professional choristers; they are usually paid something, though. These payments, along with those made for replacements in case of sickness, etc. are what cause the average figure to somewhat overstate average annual earnings of choristers. The pay agreement (*Tariffausschluss*) between the Deutscher Bühnenverein and the GDBA allows for an annual increase of two per cent per annum for all singers with a *Fest* contract (inflation rates are typically low in Germany).

As with employment, the situation with regard to soloists' pay and earnings is not so straightforward and meaningful average earnings cannot be worked out in the same way because the figure for the outlay on singers includes those on evening and guest as well as on normal contracts; as explained above, figures are only available for the number of singers with the latter. However, some information on typical payments by several opera-houses was gained from discussions with opera managements and agents. In bigger houses, soloists with a *Fest* contract earned a salary of between 3,000 and 10,000 DM a month in 1990, depending upon their status, etc., the equivalent of around £13,000–£43,000 a year; in a medium-sized house between 2,500 and 5,000 DM a month, £10,700–£21,000 a year. But this has to be regarded as their minimum pay, first, because singers get paid for performances over the number specified on their contracts in their 'home' opera-house that can add anything from 500–8,000 DM (£180 to nearly £3,000) per performance, again depending on the role, their Fach and so on. If a singer had a limit of thirty or forty performances in their contract, they might be able to do twice that number in a year by doing extra performances or guesting. Payments for guest or evening performances also vary with different houses; a house with a semi-stagione system relying on guests may pay much more than a house which operates a repertory system. Performance fees for guests, etc. quoted varied accordingly; a semi-stagione house might pay from £2,000 to £11,000 per performance; a repertory house £1,400–£3,600 in the bigger ones or £350–£1,600 in the smaller ones. The international level houses such as Munich probably pay much more but the very small houses cannot go below the agreed minimum. The contract for one evening's solo performance (*Abendvertrag*) also includes a duty to rehearse for two to three days without extra pay. But a guest singer who takes part in a new production with, say five to six weeks' rehearsal would be paid the equivalent of half his or her regular salary or one-twelfth of one performance fee per day of rehearsals (i.e. six weeks of rehearsals, six days a week are equivalent to three performance fees). Travel costs are also paid, but not subsistence.

EMPLOYMENT OUTSIDE OPERA

The only opportunities for professional choral singing in Germany outside opera are with the radio choirs; there is only a minute pool of freelance choral singers. One or two groups have developed in Early music along the lines of the British Monteverdi Choir but it is difficult for professional freelance choirs to compete with the well-established amateur choral societies on the one hand and the radio choirs on the other. There are nine radio stations in the FDR with a regular choir, each employing about forty singers. Undoubtedly, unification has increased this number but statistics were not available.

The radio stations, the choral societies, music societies, orchestras, and other promoters, particularly the city-owned concert halls, all hire soloists; the singers may be opera singers guesting or freelance concert singers, German or foreign. There is a considerable amount of employment in this field of concert and Lieder singing but, unfortunately, there was no way of quantifying it.

APPENDIX 3

AGENTS IN GERMANY

In Germany, agents are licensed to work either in the field of opera or concerts; a concert agent may arrange an evening contract in opera for a singer he or she represents for concerts but may not negotiate a normal or *Fest* contract in opera. Nor may an agent make an exclusive contract with a singer. That being said, however, agents perform more or less the same function as in Britain, which is putting singers in touch with potential employers, negotiating contracts and finding singers for opera-houses.

Practically all auditions with opera-houses are arranged through agents, unless a singer has a personal contact. Agents audition singers regularly, usually on the same day of the week, and then recommend them to opera-houses (see Legge 1988). Their job is to get the most favourable conditions in the contract, particularly with respect to the amount of free time. They get a commission of 6 per cent for a normal contract and 10 per cent for an evening contract, of which the singer pays half and the opera-house pays half.

It is tempting to conclude that the life of an agent in a labour-market with excess demand and relatively long term employment stability is a very easy one. However, since the number of agencies is rigidly controlled and only ten licences are issued, there are a lot of singers to be dealt with by a few agents; in addition, there is a great deal of circulation of singers from one house to another; there are a large number of opera-houses to be dealt with and also singers' contracts can be renegotiated as they are changed—those with growing careers wanting a better deal, and so on. Agents handle all this. They are therefore both middlemen and they perform search and information functions for both singers and opera-houses. By contrast to Britain, where the agent's commission is entirely paid by singers, at least the opera-houses make their contribution for this service.

CONCLUSIONS

It seems to be the case that many more singers are trained in Germany than in Britain but there are clearly many more employment opportunities for them. There is no unemployment among recent graduates of the music high schools—on the contrary, the problem seems to be that singers can all too readily get regular jobs, at least in small opera-houses, where their sometimes inadequate training is exposed to the public and they are in danger of singing too much too soon and may suffer vocal damage as a result. This is a sellers' market for singers and one feature of it is that singers can get used to a bureaucratically controlled routine and develop what might be called a 'welfare state mentality'. One reason for the apparently large number of foreign singers in Germany is the shortage of good German singers; another is that foreign singers, particularly Americans, are more rigorously trained and so are hired in preference to German singers.

Apart from the (universal) problem of good vocal training, standards of sight-singing are not as high as those in Britain and America so that singers, soloists, and choristers, need a lot of music coaching and rehearsals to learn parts; this is expensive in terms of the singers' and repetiteurs' time. Since many singers are on *Fest* contracts, they are not expected to learn roles in their own time; coaching is provided by the opera-house as a contractual obligation.

What Germany has which does not exist in Britain is a structure of informal on-the-job training opportunities that carries with it job security and a regular salary. Depending on where they go to work, young German singers can expect to get a two year *Normal* contract in an opera studio or more probably in a smallish opera-house, either in the choir or as soloists (or a mixture of the two) and work their way up. If they do not make a big career, they can nevertheless spend a useful working life with regular employment and a salary with virtual job security up to the age of 65. They are city or state employees and enjoy a good status in society and, so to say, do a job like anyone else. The least they would earn in 1990 was over £11,000 a year which will rise with inflation and many singers were earning over £21,000 in 1990 in choruses, and anything from £12,000 upwards as soloists. How much soloists can earn is impossible to say with any precision; in a biggish house a well-established singer could earn £40,000 on a *Fest* contract with a limit of thirty or forty performances, leaving them free to earn, say, another £40,000 in their own house by doing extra performances or even more by guesting elsewhere.

The extent to which soloists are able to pick up extra work depends first upon the limit of the performances in their contract; secondly, on how many other ensemble soloists there are in their Fach who can do the performances or roles over the period in which they need free time; thirdly, on the number of performances of opera that the 'home' house mounts; and fourthly on the relative supply of singers in the whole market in their Fach. Singers who are very much in demand as guests will probably become independent of any one opera house because they do not wish to be tied. As the price of this is having to pay all your own insurance and social security payments, only singers who have very high earnings really want to do that. A fee that was quoted as 'really high' in 1990 was 20,000 DM per performance, i.e. £7,000. A singer doing 80 performances would therefore earn £560,000 and be paid extra for rehearsals lasting more than two or three days. But they would then have to pay 25 per cent more in obligatory payments than a *Fest* contracted singer. It is easy to see why singers prefer to have one foot in each camp.

This possibly also explains why relatively few German singers spend a great deal of time abroad; they do not have the financial incentives to do so as do British or American singers. The absence of a large number of German singers on the general international market certainly does not signify that there are few of them; there are many great German singers who just rarely seem to go abroad.

The path from training to employment in Germany is much smoother than in Britain. Students have an effective choice of where and how to work. Because the major part of employment open to singers—soloists or choristers—is of a regular contracted nature it would be possible to link numbers training with employment opportunities and plan the manpower requirements of the profession. Rather as in the teaching profession in Britain, the total number of singers that will be needed at a particular future date by opera-houses can be identified and, taking account of the age structure of the profession and natural attrition, the number of new recruits needed can be calculated. There is certain long run equilibrium built into the system with its high subsidies on the one hand supporting it financially and its bureaucratically set labour requirements on the other. However, even there, guesting complicates the picture. It is far from being market-oriented but as opera is part of the fabric of life in cities all over Germany, high subsidies are not controversial, though there have been signs during the 1980s of financial retrenchment in the arts in general. From the trainee's point of view, even though there is considerable certainty about opportunites for employment and earnings, this does not completely remove risk from the singing profession in Germany; singers can still have vocal problems, ill-health, not achieve their hoped-for goals, and so on, but they do not have to cope in addition with the rough and tumble of the uncertainty of the marketplace, as do their British counterparts.

BIBLIOGRAPHY

Adler, M. (1985), 'Stardom and Talent', *American Economic Review*, 208–12.

Baker, J. (1982), *Full Circle* (Julia MacRae: London).

Barton, M., and Stewart, A. (1988), *British Music Education Yearbook* (Rhinegold: London).

Baumol, W., and Bowen, W. (1966), *The Performing Arts: The Economic Dilemma* (Twentieth Century Fund: Cambridge, Mass.).

Blaug, M. (1970), *An Introduction to the Economics of Education* (Penguin: London).

—— (1976) (ed.), *The Economics of the Arts* (Martin Robertson: London).

—— (1978), 'Why are Covent Garden Seat Prices so High?', *Journal of Cultural Economics*, 2(1): 1–20.

—— (1987), *The Economics of Education And the Education of An Economist* (Edward Elgar: Aldershot, Hants.), 3–49, 129–40.

Bridges, Lord (1964), *Report by Lord Bridges' Committee on Opera Singers* (Arts Council of Great Britain, London).

Carter, A. (1990) (ed.), *British Music Yearbook* (Rhinegold: London).

Christiansen, R. (1984), *Prima Donna: A History* (Penguin Books: London).

Clark, A., and Tarsh, J. (1987), 'How Much is a Degree Worth?', in J. Gretton, and J. Harrison (eds.), *Education and Training UK 1987* (Policy Journals: London).

Cowley, M. (1991), *The Mentor Opera Handbook* (Mentor: London).

Department of Employment (1989), *New Earnings Survey 1989* (Department of Employment: London).

—— (1990), *Employment Gazette* (July) (Department of Employment: London).

Deutscher Bühnenverein (1989), *Theaterstatistik 1987/8* (Deutscher Bühnenverein: Cologne).

Devlin, G. (1992), *Beggars' Opera* (Calouste Gulbenkian Foundation: London).

Domingo, P. (1983), *My First Forty Years* (Viking Penguin: New York).

Eckstein, J., and Feist, A. (1991), *Cultural Trends 1991* (Policy Studies Institute: London).

Ehrlich, C. (1985), *The Music Profession in Britain since the Eighteenth Century* (Clarendon: Oxford).

Equity (1990), 'It's Official! Women Get Less Work and Earn Less', *Equity Journal* (Spring).

Feist, A., and Hutchison, R. (1989), *Cultural Trends 1989*, i–iv (Policy Studies Institute: London).
—— (1990*a*), *Cultural Trends 1990* (Policy Studies Institute: London).
—— (1990*b*), *Cultural Trends in the Eighties* (Policy Studies Institute: London).
Filer, R. (1986), 'The 'Starving Artist'—Myth or Reality? Earnings of Artists in the United States', *Journal of Political Economy*, 94 (1).
Ford, T. (1986), *The Musician's Handbook* (Rhinegold: London).
GBDA (1989), *Deutsches Bühnen Jahrbuch* (Himmelheber: Hamburg).
Gowrie, Lord (1990), *Review of the London Music Conservatoires* (PCFC: London).
Graham, R., Norman, J., and Shearn, D. (1983), 'Cost Effective Opera Subsidy', *Journal of Operational Research Society*, 34 (10).
Hines, J. (1983), *Great Singers on Great Singing* (Gollancz: London).
Hutchison, R., and Feist, A. (1991), *Amateur Arts in the UK* (Policy Studies Institute: London).
Jenkins, Sir Gilmour (1965), *Making Musicians* (Calouste Gulbenkian Foundation: London).
King, B. (1989), '1987–1988 Equity Income and Employment Survey' (Equity: London).
Kloiber, R. (1985), *Handbuch der Oper* (dtv/Bärenreiter).
Legge, A. (1988), *The Art of Auditioning* (Rhinegold: London).
MacDonald, G. (1988), 'The Economics of Rising Stars', *American Economic Review*, 78: 155–66.
McMahon, W. (1987), 'Expected Rates of Returns to Education', in G. Psacaropolous (ed.) *Economics of Education Research and Studies* (Pergamon: Oxford) 187–96.
Major, N. (1987), *Joan Sutherland* (Queen Anne Press: London).
Myerscough, J. (1986), *Facts About the Arts* (Policy Studies Institute: London).
New Grove (1980), S. Sadie (ed.), The New Grove Dictionary of Music and Musicians (Macmillan: London).
Nieuwenhuis, H. (1990), *Onderzoek Opera Studios in Europa* (Ministry of Welfare, Health and Culture, Rijswijk).
Peacock, A., Shoesmith, E., and Millner, G. (1982), *Inflation and the Performed Arts* (Arts Council of Great Britain: London).
Priestley, C. (1983), *Financial Affairs and Financial Prospects of the Royal Opera Covent Garden Ltd. and the Royal Shakespeare Company* (HMSO: London).
Robinson, K. (1982), *The Arts and Higher Education* (Society for Research into Higher Education: Guildford).
Rosen, S. (1981), 'The Economics of Superstars', *American Economic Review*, 71: 845–58.

Rosselli, J. (1984), *The Opera Industry in Italy from Cimarosa to Verdi* (Cambridge University Press: Cambridge).

—— (1991), *Music and Musicians in Nineteenth Century Italy* (Batsford: London).

Santos, F. (1976), 'Risk, Uncertainty and the Performing Artist', in Blaug, (1976).

Smith, A. (1976), *An Inquiry into the Nature and Causes of the Wealth of Nations*, ed. A. Skinner and T. Wilson (Oxford University Press: London).

Throsby, D. (1986), *Occupational & Employment Characteristics of Artists* (Australia Council: North Sydney).

—— (1992), 'Artists as Workers', in Towse and Khakee (1992), 201–8.

Towse, R. (1992*a*), 'The Earnings of Singers: An Economic Analysis', in R. Towse, and A. Khakee (eds), *Cultural Economics* (Springer: Heidelberg), 209–17.

—— (1992*b*), *Economic and Social Characteristics of Artists in Wales* (Welsh Arts Council: Cardiff).

—— (1992*c*), *The Labour Market for Artists* (Exeter Discussion Papers in Economics, 92/02; University of Exeter).

—— and Khakee (1992) (eds.), *Cultural Economics* (Springer: Heidelberg).

Vaizey, J. (1978), *Training Musicians* (Calouste Gulbenkian Foundation: London).

Vishnevskaya, G. (1984), *Galina* (Sceptre: Sevenoaks, Kent).

Waits, C., and McNertney, E. (1980), 'Uncertainty and Investment in Human Capital in the Arts', in W. Hendon, J. Shanahan, and A. McDonald, (eds.), *Economic Policy for the Arts* (Abt: Cambridge: Mass.)

Willatt, H. (1976), *Report to the ACGB on the Present Facilities for Advanced Opera Training in Great Britain and Recommendations for the Future* (Arts Council of Great Britain: London).

Williams, G., and Gordon, A. (1981), 'Perceived Earnings Functions and Ex Ante Rates of Return to Post Compulsory Education in England', *Higher Education*, 10.

Withers, G. (1985), 'Artists' Subsidy of the Arts', *Australian Economic Papers* (Dec.)

INDEX

Abbey Opera 12, 45
Adler, M. 160
adult education 12, 45–6, 47, 187
age-earnings profile 152, 166–9, 226
agents 8, 11, 13, 78, 98, 100, 125, 126, 128, 133, 135, 136, 141, 142, 150, 192
　economic role of 14–15, 75, 199
　German 100, 239, 240
Aldeburgh Foundation, *see* Britten–Pears School
amateur singers 1, 2, 25, 27, 46, 47, 61, 88, 89, 109, 111, 238
amateur opera companies 12, 86, 100, 137–8
amateur performing societies, *see* choral societies
AOTOS 47, 48, **206**, **207**
Arts Council 3n, 4, 5, 27, 42, 55, 90, 116–17, 187, **215**
Associated Board exams 31, 32, 37
auditions 6, 13, 14, 15, 47, 157
　for entry to courses 30, 37, 45, 46
　in Germany 240
　as part of search process of 52, 75, 100, 199, 200
　see also search and information costs

Barton, M., and Stewart, A. 30, 31, 33, 49
Baumol, W., and Bowen, W. 107–8, 113, 115, 120
Bayreuth 3, 233
BBC 4, 9, 90–1, 97, 154
BBC Singers 4, 12, 87, 100, 129, 131, 151
Beaufort Opera 45
Birmingham Conservatoire 33, 66, **205**
Blaug, M. 112–13, 170 n., 171 n.
Bridges, Lord 27, 44
British Youth Opera 44, 132
Britten–Pears School of Advanced Musical Studies 12, 45
Buxton Festival Opera 4, 83–4

Callas, M. 17
Calouste Gulbenkian Foundation, *see* Gulbenkian Foundation
career decision 164, 173–4, 191, 202–3, 204, 232
career structure 15–16
Carter, A. 46
cathedrals, *see* Church
choir school 25, 29–30
choirs 5, 16, 25, 38–9, 40, 62, 77, 87–8, 89, 91–2, 97, 106, 109, 131, 133, 134, 156, 200, 230, 239
choral scholars 5, 12, 26, 38, 61, 77, 100
choral societies 12, 14, 90, 97, 98, 102, 111–12, 126, 133, 138–40, 145, 154, 157, 194, **211**, **216**, **217**
　see also oratorio
choristers 1, 5, 9, 18, 42, 93–6, 109, 130, 131, 136, 137
　in Germany 234, 236, 237, 238, 241, 242
chorus, *see* choir *and* opera chorus
chorus singer, *see* chorister
Christiansen, R. 1 n., 25 n.
Church, the 1, 4, 9, 12, 25, 91–2, 97, 100, 129, 131, 144, 230
City of Birmingham Touring Opera 4, 109, 138 n.
Clark, A. 170, 172, 182 n.
Clonter Opera Farm 12, 46
coaching 17, 40, 43, 54, 59, 62, 63, 77, 147, 148, 231, 241
Colchester Institute 53, 67
Colleges of Further Education 5, 26, 30–2, 47, 144
Colleges of Higher Education 5, 26, 28, 33–5, 36–41, 47, 50–3, 58, 64, 65–6, 67, 69, 72–3, 76, 144, 181, 187, 204, **206**
competitions, singing 12, 75, 92, 100, 155, 157, 177, 187 n., 199, 200, 203
concert and sessions, employment in 16, 87–9, 97, 100
　fees in 9, 15, 96, 133–4, 144, 152, 154

conditions of employment/work 6, 92–8, 103, 118, 133, 229
conservatory 2, 3, 5, 10, 26, 44, 69, 230
   see also music college
contemporary music 6, 17, 59, 87, 155
contracts 2, 3, 6, 14, 83, 87, 92–8, 118, 125, 231, 233–7, 240
cost of training singers 9, 16, 27, 54, 56, 61–2, 75–8, 163, 188, 190, 198, **208**; *see also* training, economic analysis of
costs of learning, *see* learning costs
   direct 54, 57, 62–73, 166–9, 170; *see also* Gowrie Report
   indirect 54, 57, 61, 73–5, 166–9; *see also* opportunity cost
   social 170; *see also* external costs; subsidy to training singers
   in rate-of-return calculation 225–7
   marginal, *see* marginal costs
   search and information, *see* search and information costs
covers 11, 83, 137, 141 n., 148, 232
Cowley, M. 138
Cross, J. 44

Dartington College of Arts 33, 67
Dartington Summer Music School 46
demand:
   elasticity of 105–8, 116, 117
   for labour 7, 81, 102
   for singers 5, 8, 10, 17, 81–92, 98–9, 102–5, 117, 118, 119, 120–1, 160, 161, 191, 192, 194, 201, 233
Department of Employment 129, 131, 143, 152, 172, **208**
DES 66, 170, 182, 221, 225
Deutscher Bühnenverein 229, 233, 235, 237, 238
Devlin, G. 86 n., 138 n.
discrimination, racial and sexual 7, 153, 155–7
Domingo, P. 110, 112, 141, 159
Dorset Opera 84
D'Oyly Carte Opera 4

Early music 6, 12, 16, 29, 88, 100, 109, 148, 239
earnings, singers' 19, 42, 74, 95, 119, 120, 121, 125–53, 163, 166, 172–81, 195, 197–8, 201, 202, **218–19**, **220**, **221**, **222**, **223**

distribution of 158–62, 175, 177, 192, 195, 208
   in Germany 238–9
   in rate-of-return calculation 225–8
   see also concert and sessions; opera; oratorio; orchestras
earnings forgone 56, 57, 60, 73–4, 77, 164, 168–9, 198, **222**, **223**, 225–6
   see also opportunity cost
Eckstein, J. 209
Ehrlich, C. 3 n.
employers of singers 6, 8, 11, 13, 18, 23, 24, 34, 52, 55, 56, 57, 58, 100, 102–4, 126, 147, 155, 192, 194, 199, 240
employment of singers 8–9, 27, 39, 53, 81–92, 95, 100, 102, 164, 175, 176, 192, 193, 197, 204
   in Germany 229, 230, 233, 237–8, 239, 240, 241, 242
English National Opera 4, 11, 82–3, 94, 96, 114–16, 118, 129, 130, 159, **210**, **212**
Equity 9, 10, 15, 88, 93, 96, 97, 102, 118, 126, 127–8, 130, 131, 132, 133, 134, 138, 147 n., 150, 151, 154, 156, 192, 194
excess supply 8, 78, 105, 178–9, 196, 199
   of labour 7, 23, 61, 102–5, 197
   of singers 5, 6, 8, 9, 10, 18, 23, 35, 52, 54, 75–8, 160, 191, 192, 193, 197, 202, 241
   see also oversupply
expenses, singers' 128, 137, 140, 144, 145–50, **220**, **221**, **222**, 226
external costs and benefits 184–6

Fach 17, 234–5, 236, 239, 241
fees in higher education 9, 42, 65, 67–8, 70, 73, 169, 181, 192, 195, 200, 202, **207**
fees of singers 1, 2, 6, 7, 8, 9, 10, 15, 16, 18, 59, 74–5, 96, 98, 105, 111, 112, 114, 116, 119, 125, 126, 131, 133, 134–41, 142, 144, 146, 149, 150, 160, 161, 162, 172, **218–19**, **220**
   in Germany 235, 239, 241
   as a signal 135, 160–1, 162, 200
   see also concert and sessions; opera; opera choruses; oratorio
fees of singing teachers 46, 49, 67, 145
Feist, A. 128, 138 n., 209, **210**, 215

INDEX

Filer, R. 153, 165
Finch, H. 43
fixer 6, 39, 52, 61, 87–8, 134, 194, 199–200
Ford, T. 147 n.
freelance work 3, 4, 5, 8, 15, 24, 55, 58, 74, 75, 81, 87, 93, 125, 132–4, 149, 150, 151
  in Germany 231, 232, 236, 237, 238, 239

GBDA 229, 233, 238
Germany 3, 17, 44, 100, 119, 195–6, 229–42
Glyndebourne Festival Opera 3, 4, 5, 11, 42, 83–4, 99, 120, 132, 157
Glyndebourne Touring Opera 4, 5, 11, 83–4, 132, 138, **210**
Gordon, A. 163
Gowrie Report 66, 68–73, 76, 77 n., 172, **208**, 226
Guildhall School of Music and Drama 30, 33, 47, 66–7, **205**, **207**
Gulbenkian Foundation 13 n., 23, 27, 86 n.

higher education 5, 23, 24, 25, 26, 27, 28, 33–41, 43, 47, 50–3, 54, 55, 58, 62, 73, 77, 92, 145, 164, 168, 169, 181–3, 186, 187, 189, 192, **209**, 228
  funding of 9, 63–6
  screening function of 52, 199, 203
Hines, J. 25 n.
Huddersfield Choral Society 90
Huddersfield School of Music 67
Hutchison, R. 128, 138 n., **210**, **215**

incentives, economic or financial 7, 10, 43, 55, 56, 58, 60, 65, 67–8, 98, 116, 162, 163, 196, 197, 200, 203, 236, 241
inflation 9, 63, 129, 152, **214**, 238, 241
information, economic role of 14, 15, 177
  cost of, *see* search and information costs
investment in human capital 60, 163, 164–6, 168, 175, 179, 228
ISM 9, 47, 93, 97, 98, 126, 128, 129, 139, 140, 141, 144, 146, 150, 151, 154, 156, 192
Italy 1, 3, 119, 159, 232

Jenkins, G. 27, 68

Kent Opera 4, 84, 99, **210**
Kentish opera 84
King, B. 127
Kloiber, R. 235

labour-market 6–10, 52, 58, 76, 125, 126, 153, 191, 195, 196, 199
  for singers 3, 19, 26, 27, 28, 35, 41, 51, 52, 53, 75, 81, 87, 135, 155, 188, 191–6, 197, 200, 204, 237, 240
  segmentation 7, 153
  see also market for singers
learning costs 55, 60, 93, 95, 118, 147–8, 149, 241
Leeds College of Music, City of 33
Legge, A. 13, 17, 232, 235, 240
Lind, J. 25
Local Authority Music Centres 26, 30
London Chamber Opera 85
London College of Music 33, 66, 67
London Opera Centre 27, 42, 50
London Opera Players 85

MacDonald, G. 160, 161, 177, 179
McMahon, W. 171 n.
McNertney, E. 179, 180
Major, N. 141
Malibran, M. 25
manpower planning 23–5, 194, 195, 203, 242
marginal costs in music colleges 64–5, 67–8, 70, 71
market for singers 1, 2, 34, 81, 82, 92–3, 99, 102–5, 119, 121, 126, 135, 153–8, 188, 192, 199, 201, 240
  organization of in Britain 10–16, 81–98
  international 2, 18–19, 58, 140, 202, 240, 241
  specialization in 16–18, 60, 89, 119
masterclasses 12, 45, 49, 51, 62, 63, 92
Mayer–Lismann Opera Workshop 45
median fee, payment or earnings 128, 139, 140, 144, 146, 149, 150, 158, 179, 197, **216**, **217**
Millner, G. 113–16, 152, **213**, **214**
Monteverdi Choir 87, 109, 239
Morley College 12, 45
Music at Oxenfoord 46

music colleges 3, 5, 10, 12–14, 16,
    25–6, 33, 34, 47, 50–3, 55, 61, 87,
    172, 177, 187, 204
  costs in 67–73, 76, 77, 170, 186,
    198
  course content in 37–40, 58, 62, 63
  funding of 64, 66–7, 181
  number of places for singers in 9,
    24, 28, 68, 69, 195, 201
  output of singers from 23, 35–6,
    205, **207**
  teachers' pay in 144
music high schools (Hochschule für
    Musik) 230, 240
musicals 9, 11, 59, 86–7, 97, 118, 127,
    130–1, 137 n., 144, 151, 154,
    in Germany 233, 236

NAB 181
National Opera Studio 5, 11, 13, 14,
    27, 42–4, 50, 51, 55, 58 n., 62, 71,
    74, 77, 100, 231
New Earnings Survey 74
New Sadlers' Wells Opera 4
NFMS 90, 97, 138, 211, 216, 217
Nieuwenhuis, H. 44

opera choruses 11, 15, 40, 55
  cost of 118
  earnings in 114–15, 129–30, 146,
    149, 151, 152, 154, 174, **212**
  employment in 82–4, 86, 99
  in Germany 231, 233–4, 239, 241
opera 1, 3, 4, 5, 8, 9, 16, 17, 28, 39,
    40, 43, 71, 81, 89, 97, 105, 106,
    107, 110, **210**
  costs of 112–16, 118–20, **214**
  earnings in 8, 9, 127–32, 136–8, 143,
    119, **210** 19
  employment in 13, 18, 82–7, 93–6,
    141, 142, 156, 157, 194
  in Germany 3, 229, 231–9, 240
  subsidy of 3, 102, 116–20, **215**
  training in 31, 39–40, 41–6, 55, 59
Opera East 109
Opera 80 4, 83–4, 109, 132, 138
Opera Factory 83–4, 132
Opera Integra 45
Opera North 4, 11, 82–3, 210
Opera Northern Ireland 4, 83–4
Opera Viva 45
opportunity cost 56–7, 61, 73–5,
    147–8, 160, 168–9, 170, 182, 189

oratorio, employment in 3, 4, 5, 89,
    90, 141, 142
  fees in 135, 138–40, **218–19**
  see also choral societies
orchestras, employment with 9, 87,
    89–90, 98, 102
  fees paid by 133, 134–5, 140
oversupply of singers 8, 77, 120, 156,
    180, 196–202, 203–4

Pavarotti, L. 176
Pavilion Opera 4, 84–5, 109,
pay/payment, rate of 6, 8, 10, 93, 94,
    96, 106, 115, 120, 125, 130, 131,
    137, 191, 192, 193, 196–7
  in Germany 238–9
  minimum 9, 10, 93, 96–8, 102–4,
    126, 132, 154, 161, 191
PCFC 65, 66–7, 68–73, 76, 181, 182,
    209
Peacock, A. 113–16, 152, **213**, **214**
perfect pitch 17, 55
polytechnics 5, 23, 26, 28, 33–5,
    36–41, 50–3, 58, 63, 64, 65–6, 67,
    69, 72–3, 76–7, 144, 181, 187, 198,
    204, 206
postgraduate training, *see* training,
    postgraduate
price of tickets 106–7, 112–13, 116,
    117, 120, 159, 192, 194
Priestley, C. 104
principals, opera 9, 15, 83, 109, 112,
    114, 115, 118, 119, 129, 130–1,
    136, 137, 156, **212**
professional singers 1, 2, 4, 5, 27, 29,
    61, 63, 84, 85, 86, 89, 90, 131, 136,
    137, 138, 140
psychic benefits, income 164, 165, 180,
    182, 189, 197
Purcell Room 91, 107, 209

Queen Elizabeth Hall 107, 209

radio 9, 88, 96, 97, 239
rate of return 61, 161–74, 178, 181–3,
    186, 187–8, 189, 190, 196, 197,
    198, 199, 203
  calculation of **222**, 224–8
recitals 5, 16, 17, 40, 90–1, 106, 141,
    144, 148
recording 9, 96, 105, 126, 133–4, 141,
    201–2, **218**
repertory system 232–3, 236, 239

INDEX

repetiteur, *see* coaching
Ricciarelli, K. 106, 112
risk 7, 91 n., 174–81, 188, 189, 195, 196, 201, 202, 204, 242
Robinson, K. 28, 34, 50
Rosen, S. 159–60, 161, 175, 179
Rosselli, J. 1 n., 14, 135, 232
Royal Academy of Music 33, 66, 67, 68–73, **205**, **207**, **208**
Royal Choral Society 90
Royal College of Music 33, 66, 67, 68–73, **205**, **207**
Royal Festival Hall 107, 209
Royal Northern College of Music 33, 39, 66, 71, **205**, **207**
Royal Opera (Covent Garden) 3, 4, 11, 82–3, 94, 96, 112–13, 114–16, 129, 136, 137, 159, **210**, **212**
Royal Scottish Academy of Music and Drama 33, 66, 67, 71, **205**, **207**

salaried singers 1, 125, 126, 129–32
  in Germany 3, 235, 241
Santos, P. 180
schools 25, 26, 28, 29–30, 49
Scottish Opera 4, 11, 82–3, 107, 114–16, **210**, **213**
search and information costs 15, 75, 99–100, 144, 160, 162, 199–200, 202, 240
semi-professional singers 2, 46, 84, 85, 86, 136
sessions singers, *see* concert and sessions
Shoesmith, E. 113–16, 152, **213**, **214**
sight-singing 12, 16–17, 39, 87, 155, 241
singing teachers 92, 177, 193 n.
  fees and earnings of 62, 67, 98, 126, 144–5
  private 5, 26, 46–50, **206**, **207**
Smith, A. 1–2, 176, 178–9
soloists 1, 11, 12, 26, 39, 77, 82, 154
  employment for 86, 87, 88, 89, 90–1, 109, 156
  fees of 97, 98, 111, 128, 134–41, 144, 146, **211**, **216**, **217**, **218**, **219**, **221**
  in Germany 231, 234, 235, 236, 237, 238, 239, 241, 242
  see also principals, opera
stagione system 110, 232, 233, 236, 239

Studio, *see* National Opera Studio
subsidy:
  to the arts 3, 5, 9, 24, 58 n., 102, 116–20, 159, 183–6, 191, 192, 196, 204, **215**, 242
  to education 5, 54, 56, 57, 78, 169, 171, 180, 181, 182, 183–6, 192, 204
  to training singers 68–71, 72–3, 74, 186–90, 191, 199, 203–4, 227
  see also higher education, funding of
substitution 17–18, 102, 106, 107–11, 112, 113, 116, 117, 118, 120, 121, 154, 155
Summer Music 46
superstars 142, 143, 154, 159–62, 179, 199–202
supply, excess, *see* excess supply
Sutherland, J. 141

Tarsh, J. 170, 172, 182 n.
taxes 132, 147–50, 185, 198, 202, 238
Throsby, D. 165 n.
Towse, R. 6, 161, 165 n.
trade unions 6, 7, 9, 102, 103–4, 126, 229
trained singers, number of 31–2, 33, 35–6, 42, 43, 44, 45, 48, 49, 50
training, economic analysis of 54–62
  see also rate of return
training of singers 1, 5, 9, 16–17, 24, 25–9, 37–41, 55, 60, 62, 86, 164
  advanced 5, 27, 41–6, 54, 77, **207**, **208**
  formal 10, 11, 26, 34, 40, 54, 57, 61, 62–74, 163, 168, 170, 204
  in Germany 230–1, 241, 242
  on-the-job 12, 28, 40, 42, 44, 57, 61, 62, 71, 83, 85, 168, 204
  postgraduate 5, 10, 28, 34, 40, 41–2, 49, 54, 77, **205**, **207**
  private 5, 30, 35, 38, 46–50, 55, 61, 62–3, 164, 192, 193 n.
Travelling Opera 4, 84–5, 109, 132
Trinity College of Music 33, 47, 66, 67, 68–9, **205**, **207**
TV 4, 9, 88, 96, 97, 126, 133, 201, 202

UFC 65, 76, 181, 182, **209**
uncertainty, *see* risk
unemployment of singers 8, 9, 105, 127–8, 153, 154, 175, 193, 197, 198, 240

universities 5, 23, 26, 28, 32, 33–5, 36–41, 50–3, 58, 64–6, 72, 76, 144, 170, 181, 186, 187 n., 198, 204, **206**

Vaizey, J. 13 n., 27–8, 29, 34, 50–1, 68, 76
Vishnevskaya, G. 196 n.
vocational training 34, 40–1, 43, 52, 57–8, 62, 76

wage, wage rate 1, 6, 7, 9, 93, 114, 116, 125, 153, 161, 195

Waits, C. 179, 180
Welsh College of Music and Drama 33, 66
Welsh National Opera 4, 11, 82–3, 114–16, 159, **210**, 213
Wigmore Hall 91, 187 n., **209**
Willatt, H. 27–8, 35, 43, 44
Williams, G. 163
Withers, G. 165
Wood, A. 44

Young Concert Artists' Trust 14